1998 YEARBOOK of ASTRONOMY

1998 YEARBOOK of ASTRONOMY

edited by

Patrick Moore

MACMILLAN

First published 1997 by Macmillan
an imprint of Macmillan Publishers Ltd
25 Eccleston Place, London SW1W 9NF
and Basingstoke

Associated companies throughout the world

ISBN 0 333 67537 1

9 8 7 6 5 4 3 2 1

A CIP catalogue record for this book is available from the British Library.

Photoset by Rowland Phototypesetting Ltd,
Bury St Edmunds, Suffolk
Printed and bound in Great Britain by
Mackays of Chatham plc, Chatham, Kent

Contents

Editor's Foreword . 7

Stop Press: Pathfinder to Mars . 8

Preface . 13

Part I: *Monthly Charts and Astronomical Phenomena*

Notes on the Star Charts . 15

Northern Star Charts . 18

Southern Star Charts . 44

The Planets and the Ecliptic . 70

Phases of the Moon, 1998 . 74

Longitudes of the Sun, Moon and Planets in 1998 . 75

Events in 1998 . 77

Monthly Notes . 78

Eclipses in 1998 . 98

Occultations in 1998 . 99

Comets in 1998 . 100

Minor Planets in 1998 . 101

Meteors in 1998 . 104

Some Events in 1999 . 105

Part II: *Article Section*

Astronomers – and Squirrels
Patrick Moore . 106

Life on Mars?
Richard L. S. Taylor . 110

Galileo – A Year Among Jupiter's Satellites
David A. Rothery . 123

Cassini/Huygens – Mission to Saturn and its moon Titan
Paul Murdin . 139

Beyond Neptune – The Edge of the Solar System
Andrew J. Hollis . 157

The Enigma of Tektites
Joe McCall . 165

Dust Rings around Normal Stars
Helen J. Walker . 187

On the Brink of Black Holes – The Story of Neutron Stars
Chris Kitchin . 194

Experiments in Visual Supernova Hunting with a Large Telescope
Robert Evans . 207

The End of the Universe
Iain Nicolson . 220

Part III: *Miscellaneous*

Some Interesting Variable Stars . 233

Mira Stars: Maxima 1998 . 239

Some Interesting Double Stars . 241

Some Interesting Nebulæ, Clusters and Galaxies . 245

Our Contributors . 246

Astronomical Societies in the British Isles . 248

Editor's Foreword

The new *Yearbook* follows the usual pattern. Gordon Taylor has provided the data for the monthly charts, as he has done for so many years now; John Isles has updated the variable star notes, and Robert Argyle has dealt similarly with the double star notes.

We have a varied selection of articles, some from regular contributors such as Iain Nicolson and Chris Kitchin, and others from writers who are new to the *Yearbook*; we welcome Drs Taylor, Rothery, Walker and McCall. We also have a contribution from the Revd Robert Evans, the Australian clergyman who has made such important contributions to the study of supernovæ and whose reputation is worldwide.

Unfortunately, it is now necessary to complete the *Yearbook* earlier than used to be the case, because of revised publication schedules, and this means that we cannot give full news about subjects such as Comet Hale–Bopp and the various Martian probes. We will, however, return to them in the 1999 Yearbook.

PATRICK MOORE
Selsey, December 1996

Stop Press:

Pathfinder To Mars

On 4 July 1997 – America's Independence Day – the space-probe Pathfinder touched down on the surface of Mars. It had been on its way since 4 December 1996 and, on arrival, its distance from Earth was over 120,000,000 miles, so that a radio signal would take between ten and eleven minutes to get there.

The last successful landings on Mars date back to 1976, when Vikings 1 and 2 sent data back direct from the surface and also searched – without result – for any trace of Martian life. The landings had been 'soft', and the probes were brought down by rocket braking. Pathfinder was different. This time the space-craft was encased in airbags, and the plan was to come down quite violently, so that the whole craft would bounce several times before coming to rest. The first bounce would carry it to a height as great as that of a six-storey building. On arrival, the airbags would be deflated; the craft would 'stand upright', and from it a small rover, the Sojourner, would crawl down a 35-degree ramp and start exploring the Martian surface. It was a totally new technique, and nobody was entirely confident that it would work. In fact it did, and before long the first images were being received by Mission Control. It can well be appreciated that the mood there was nothing short of ecstatic.

The selected site was Ares Vallis, the mouth of an ancient channel which had almost certainly been cut by water millions of years ago; there can be no open water on Mars now, because the air-pressure is too low (below 10 millibars everywhere), but at least there is plenty of ice, so that in this respect Mars will help future colonists much more than the Moon can do. Ares Vallis must once have been a raging torrent, bringing rocks of all kinds down to its mouth and flooding the adjacent plain, so that it was expected that the materials available for study would be very varied. This indeed proved to be the case.

The first pictures showed a red, rock-strewn landscape under a brownish sky; there is too little air on Mars to make the sky blue, but too much to leave it black. The overall hue is due to dust suspended in the atmosphere, and certainly Mars is a 'dusty' place. The general view was rather different from the scenes sent back by the earlier probes; Viking 1 had come down in the 'golden plain' of Chryse, while Viking 2 landed on the more northerly plain, Utopia. Ares Vallis is a very prominent feature, nearly 600 miles long, lying to the north of the dark feature known to all observers of Mars as the Margaritifer Sinus. Note that Martian valleys are not identical with the famous (or infamous) canals reported by Percival, Lowell and his colleagues earlier in our century; these canals were simply, tricks of the eye, but the old riverbeds are evident enough, and we can even see what must once have been islands.

Sojourner itself is 26 inches long and 7 inches tall; each of its wheels is 5 inches high. It weighs around 22 pounds, so that all in all it is comparable in size to a domestic television set. It is not a fast mover; on average it crawls along at 1 foot 4 inches per minute so that during its active lifetime it cannot move very far away from the main station, which has now been officially named in honour of the famous American astronomer, Carl Sagan. The camera on the main station is mounted on a retractable mast, and can extend to a total height of 5 feet 6 inches, so that it can provide a panoramic view of the surrounding area.

Local features have already been given provisional names. On the horizon we see Twin Peaks, two hills around a mile from the station and each several hundred yards high. There are some notable rocks: one was promptly nicknamed Barnacle Bill, because it gives the impression of being encrusted with barnacles, and another received the nickname, Yogi, because in profile it looks a little like a bear. Another hill was distinguished by a white streak, and was dubbed the Ski Run. To quote Dr Matt Golombek, Project Scientist for the mission: 'We have a grab bag suite of rocks here. There are different colours, textures, fabrics, sizes, shapes, which are completely different from those at the Viking sites.' There can no longer be the slightest doubt that water once roared down the Ares Vallis; in those far-off times Mars was a very different place from the dusty world of today.

But what about the all-important question of life?

Let it be stressed at once that the Pathfinder mission was not, like Viking, designed to make a search for living organisms. It was not equipped to do so, and all in all its main rôle was to demonstrate that the airbag technique was workable and to make sure that a small rover could be 'guided' around, examining rocks and sending back the results. There were many new improvements, too, and in some

ways Sojourner almost seemed to be intelligent, as it was able to avoid obstacles in its path.

Sojourner is equipped with the latest devices. There is, for instance, an X-ray spectrometer, which bombards samples with alpha-particles and then detects backscattered particles, protons and X-rays. Though its maximum range is no more than 550 yards – less than half a mile – it has plenty of samples to investigate. Originally its active lifetime was to be confined to less than seven Martian days or sols, but before long the planners announced that they hoped to continue work for rather longer than that. When it comes to the end of its career, it will simply stay on Mars until a future expedition collects it. No doubt it will end up as the prime exhibit in a Martian museum.

The triumph of Pathfinder has made up for the earlier losses of America's Mars Observer and the Russian Mars '97. Moreover another probe, Mars Global Surveyor, was on its way even as Pathfinder bounced down, and is due to reach its target in September, so that it will have arrived by the time that this *Yearbook* is in your hands. It is not a lander, but it will orbit the planet and send back maps of the surface which will – we hope – be far better than anything previously obtained.

These missions are only a start. Within ten years it should be possible to send up a sample-and-return probe, to scoop up Martian material and bring it home for analysis. Then, surely, we will know for certain whether there is, or ever has been, life on Mars. All this will be possible because of the work now going on, and there can be no doubt that Pathfinder will be remembered as one of the most brilliantly successful missions of the end of the twentieth century.

PATRICK MOORE
Selsey, July 1997

Preface

New readers will find that all the information in this *Yearbook* is given in diagrammatic or descriptive form; the positions of the planets may easily be found from the specially designed star charts, while the monthly notes describe the movements of the planets and give details of other astronomical phenomena visible in both the northern and southern hemispheres. Two sets of star charts are provided. The **Northern Charts** (pp. 14 to 39) are designed for use at latitude 52°N, but may be used without alteration throughout the British Isles, and (except in the case of eclipses and occultations) in other countries of similar northerly latitude. The **Southern Charts** (pp. 40 to 65) are drawn for latitude 35°S, and are suitable for use in South Africa, Australia and New Zealand, and other locations in approximately the same southerly latitude. The reader who needs more detailed information will find *Norton's Star Atlas* an invaluable guide, while more precise positions of the planets and their satellites, together with predictions of occultations, meteor showers and periodic comets, may be found in the *Handbook* of the British Astronomical Association. Readers will also find details of forthcoming events given in the American magazine *Sky & Telescope* and the British monthly *Modern Astronomer*.

Important Note

The times given on the star charts and in the Monthly Notes are generally given as local times, using the 24-hour clock, the day beginning at midnight. All the dates, and the times of a few events (e.g. eclipses), are given in Greenwich Mean Time (GMT), which is related to local time by the formula

Local Mean Time = GMT − west longitude

In practice, small differences in longitude are ignored, and the observer will use local clock time, which will be the appropriate Standard (or Zone) Time. As the formula indicates, places in west longitude will have a Standard Time slow on GMT, while places in east longitude will have a Standard Time fast on GMT. As examples we have:

Standard Time in

New Zealand	GMT	+	12 hours
Victoria; NSW	GMT	+	10 hours
Western Australia	GMT	+	8 hours
South Africa	GMT	+	2 hours
British Isles	GMT		
Eastern ST	GMT	–	5 hours
Central ST	GMT	–	6 hours, etc.

If Summer Time is in use, the clocks will have been advanced by one hour, and this hour must be subtracted from the clock time to give Standard Time.

Notes on the Star Charts

The stars, together with the Sun, Moon and planets, seem to be set on the surface of the celestial sphere, which appears to rotate about the Earth from east to west. Since it is impossible to represent a curved surface accurately on a plane, any kind of star map is bound to contain some form of distortion. But it is well known that the eye can endure some kinds of distortion better than others, and it is particularly true that the eye is most sensitive to deviations from the vertical and horizontal. For this reason the star charts given in this volume have been designed to give a true representation of vertical and horizontal lines, whatever may be the resulting distortion in the shape of a constellation figure. It will be found that the amount of distortion is, in general, quite small, and is only obvious in the case of large constellations such as Leo and Pegasus, when these appear at the top of a chart and so are elongated sideways.

The charts show all stars down to the fourth magnitude, together with a number of fainter stars which are necessary to define the shapes of constellations. There is no standard system for representing the outlines of the constellations, and triangles and other simple figures have been used to give outlines which are easy to follow with the naked eye. The names of the constellations are given, together with the proper names of the brighter stars. The apparent magnitudes of the stars are indicated roughly by using four different sizes of dots, the larger dots representing the brighter stars.

The two sets of star charts are similar in design. At each opening there is a group of four charts which give a complete coverage of the sky up to an altitude of 62½°; there are twelve such groups to cover the entire year. In the **Northern Charts** (for 52°N) the upper two charts show the southern sky, south being at the centre and east on the left. The coverage is from 10° north of east (top left) to 10° north of west (top right). The two lower charts show the northern sky from 10° south of west (lower left) to 10° south of east (lower right). There is thus an overlap east and west.

Conversely, in the **Southern Charts** (for 35°S) the upper two charts show the northern sky, with north at the centre and east on the right. The two lower charts show the southern sky, with south at the centre and east on the left. The coverage and overlap is the same on both sets of charts.

Because the sidereal day is shorter than the solar day, the stars appear to rise and set about four minutes earlier each day, and this amounts to two hours in a month. Hence the twelve groups of charts in each set are sufficient to give the appearance of the sky throughout the day at intervals of two hours, or at the same time of night at monthly intervals throughout the year. The actual range of dates and times when the stars on the charts are visible is indicated at the top of each page. Each group is numbered in bold type, and the number to be used for any given month and time may be found from the following table:

Local Time	18^h	20^h	22^h	0^h	2^h	4^h	6^h
January	11	12	1	2	3	4	5
February	12	1	2	3	4	5	6
March	1	2	3	4	5	6	7
April	2	3	4	5	6	7	8
May	3	4	5	6	7	8	9
June	4	5	6	7	8	9	10
July	5	6	7	8	9	10	11
August	6	7	8	9	10	11	12
September	7	8	9	10	11	12	1
October	8	9	10	11	12	1	2
November	9	10	11	12	1	2	3
December	10	11	12	1	2	3	4

The charts are drawn to scale, the horizontal measurements, marked at every 10°, giving the azimuths (or true bearings) measured from the north round through east (90°), south (180°) and west (270°). The vertical measurements, similarly marked, give the altitudes of the stars up to 62½°. Estimates of altitude and azimuth made from these charts will necessarily be mere approximations, since no observer will be exactly at the particular latitude, or at the stated time, but they will serve for the identification of stars and planets.

The ecliptic is drawn as a broken line on which longitude is marked every 10°; the positions of the planets are then easily found by reference to the table on p. 71. It will be noticed that on the Southern Charts the **ecliptic** may reach an altitude in excess of 62½°

on star charts 5 to 9. The continuations of the broken line will be found on the charts of overhead stars.

There is a curious illusion that stars at an altitude of 60° or more are actually overhead, and beginners may often feel that they are leaning over backwards in trying to see them. These overhead stars are given separately on the pages immediately following the main star charts. The entire year is covered at one opening, each of the four maps showing the overhead stars at times which correspond to those for three of the main star charts. The position of the zenith is indicated by a cross, and this cross marks the centre of a circle which is 35° from the zenith; there is thus a small overlap with the main charts.

The broken line leading from the north (on the Northern Charts) or from the south (on the Southern Charts) is numbered to indicate the corresponding main chart. Thus on p. 38 the N–S line numbered 6 is to be regarded as an extension of the centre (south) line of chart 6 on pp. 24 and 25, and at the top of these pages are printed the dates and times which are appropriate. Similarly, on p. 65, the S–N line numbered 10 connects with the north line of the upper charts on pp. 58 and 59.

The overhead stars are plotted as maps on a conical projection, and the scale is rather smaller than that of the main charts.

1L

October 6 at 5^h	October 21 at 4^h
November 6 at 3^h	November 21 at 2^h
December 6 at 1^h	December 21 at midnight
January 6 at 23^h	January 21 at 22^h
February 6 at 21^h	February 21 at 20^h

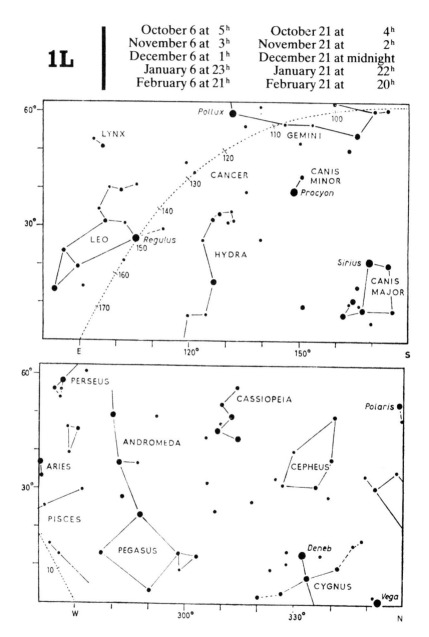

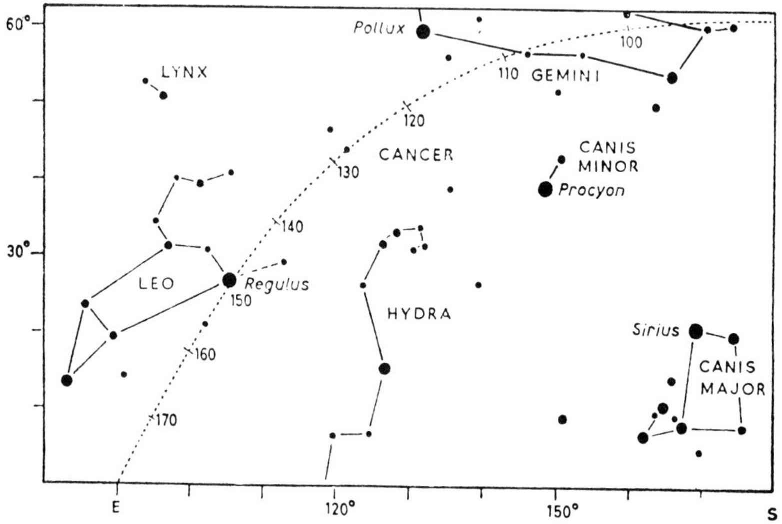

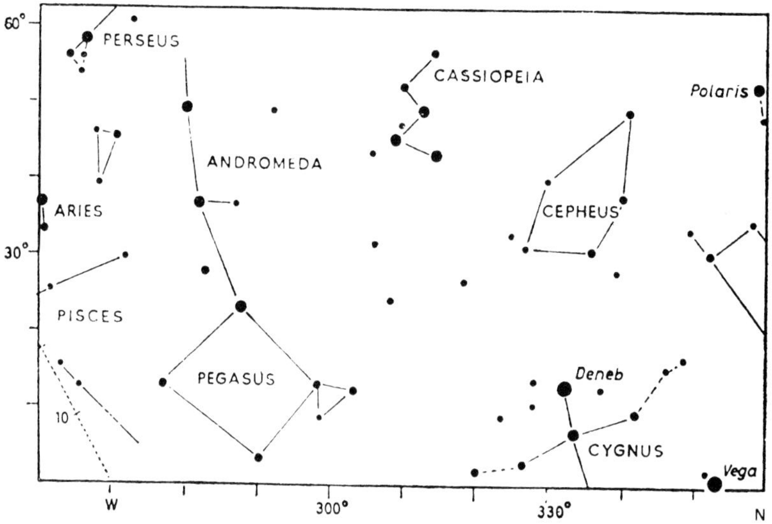

October 6 at 5h	October 21 at 4h
November 6 at 3h	November 21 at 2h
December 6 at 1h	December 21 at midnight
January 6 at 23h	January 21 at 22h
February 6 at 21h	February 21 at 20h

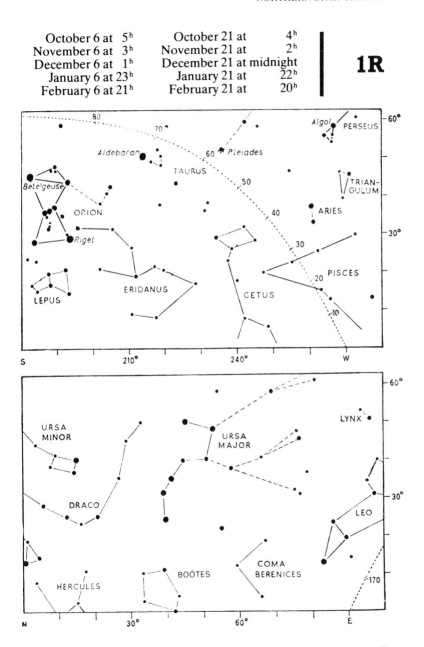

1R

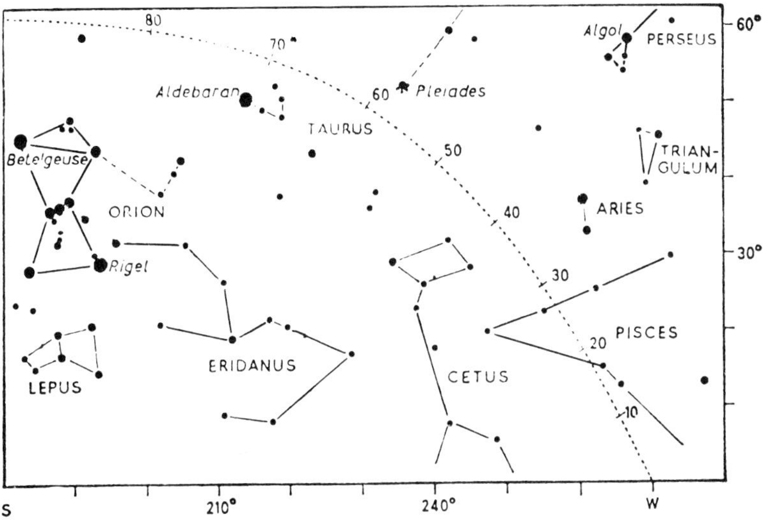

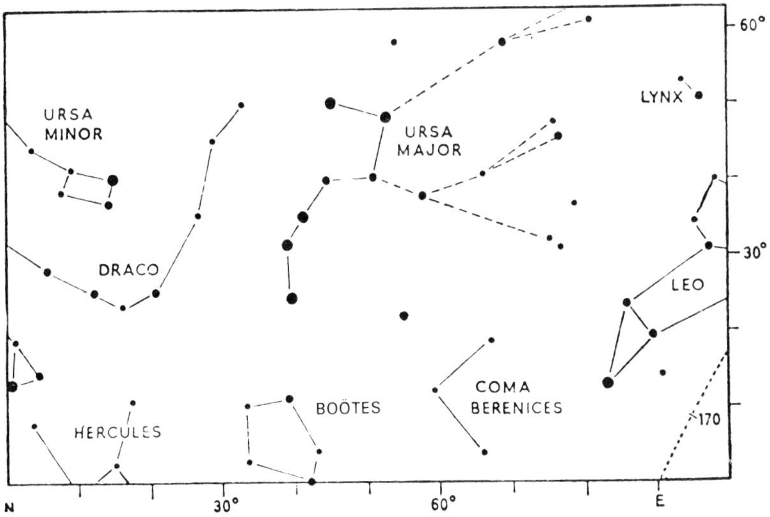

2L

November 6 at 5^h	November 21 at 4^h
December 6 at 3^h	December 21 at 2^h
January 6 at 1^h	January 21 at midnight
February 6 at 23^h	February 21 at 22^h
March 6 at 21^h	March 21 at 20^h

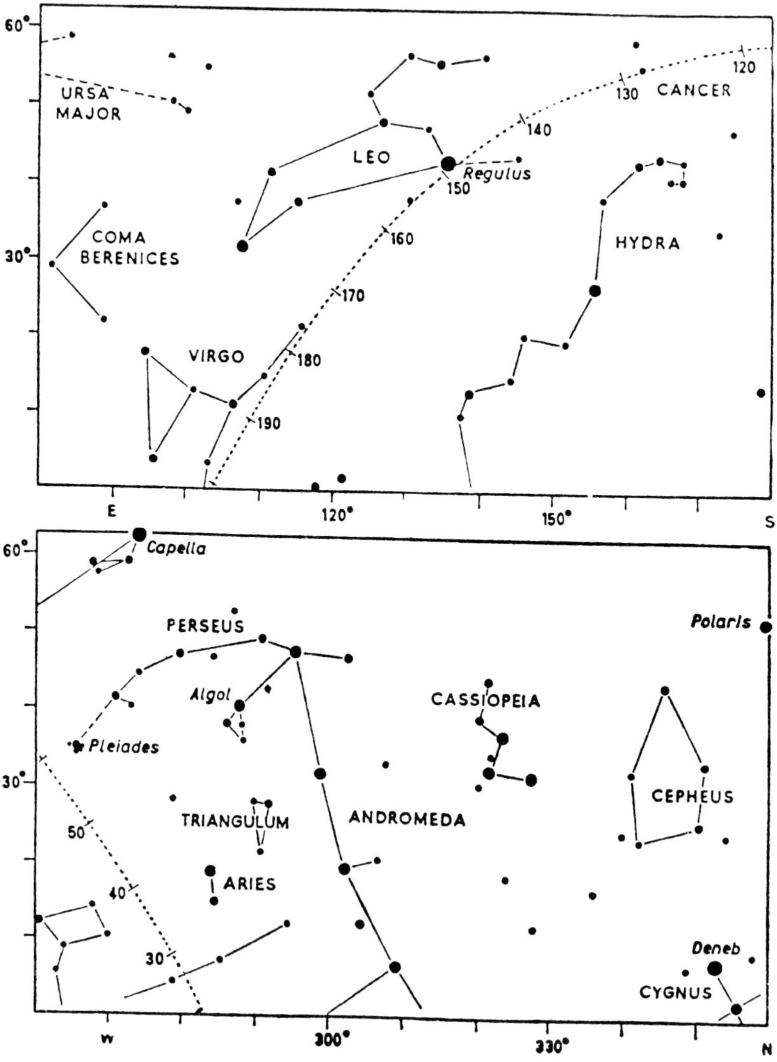

November 6 at 5^h	November 21 at 4^h
December 6 at 3^h	December 21 at 2^h
January 6 at 1^h	January 21 at midnight
February 6 at 23^h	February 21 at 22^h
March 6 at 21^h	March 21 at 20^h

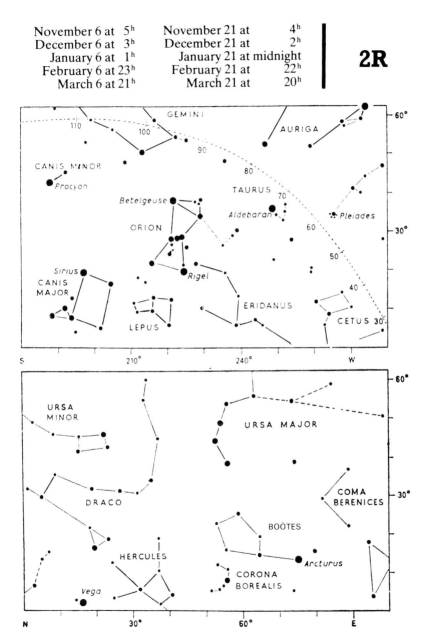

2R

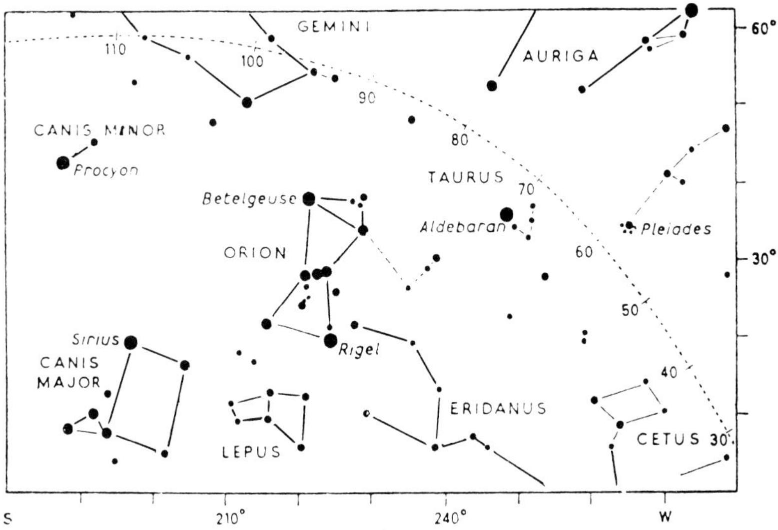

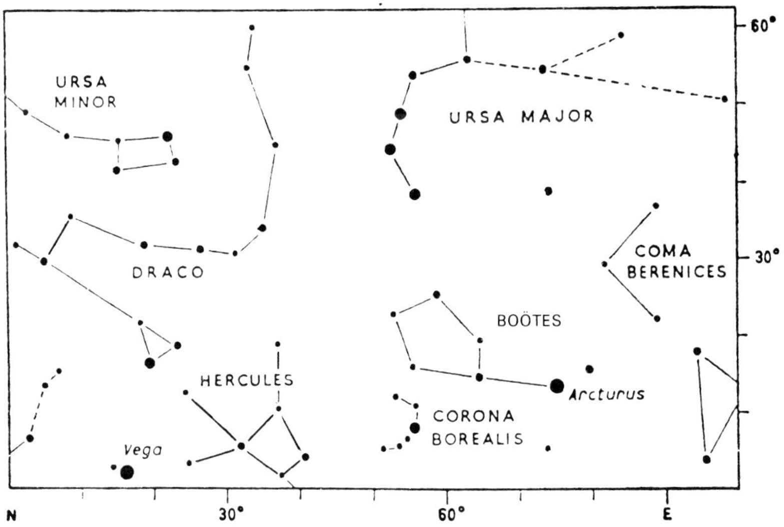

3L

December 6 at 5^h	December 21 at 4^h
January 6 at 3^h	January 21 at 2^h
February 6 at 1^h	February 21 at midnight
March 6 at 23^h	March 21 at 22^h
April 6 at 21^h	April 21 at 20^h

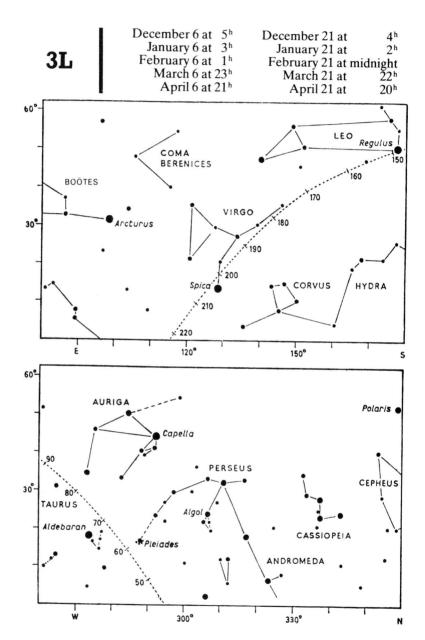

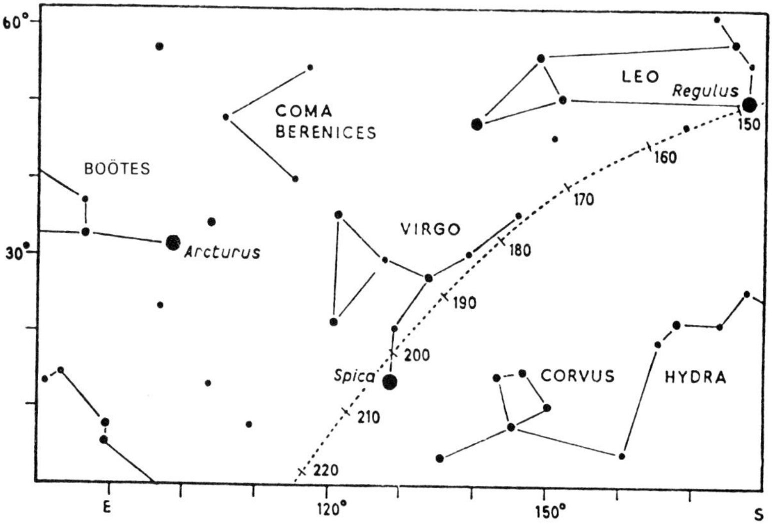

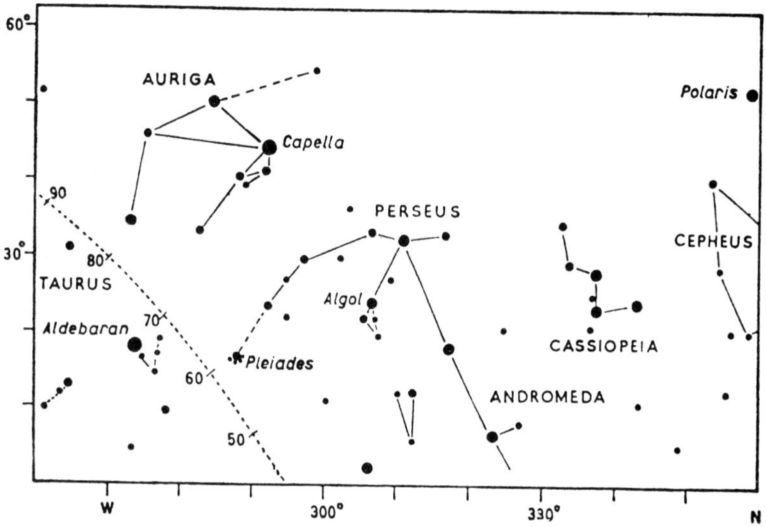

December 6 at 5h	December 21 at 4h
January 6 at 3h	January 21 at 2h
February 6 at 1h	February 21 at midnight
March 6 at 23h	March 21 at 22h
April 6 at 21h	April 21 at 20h

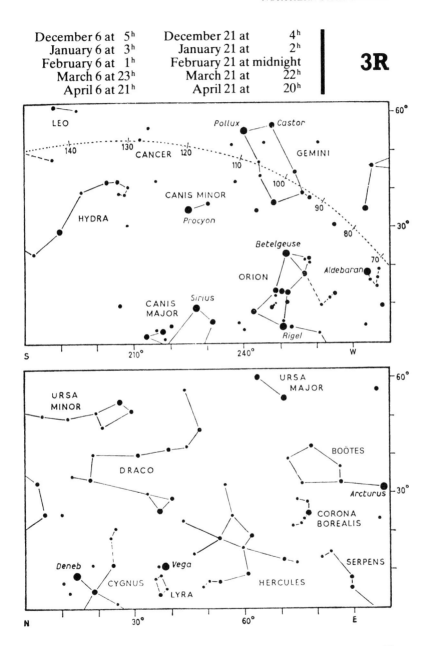

3R

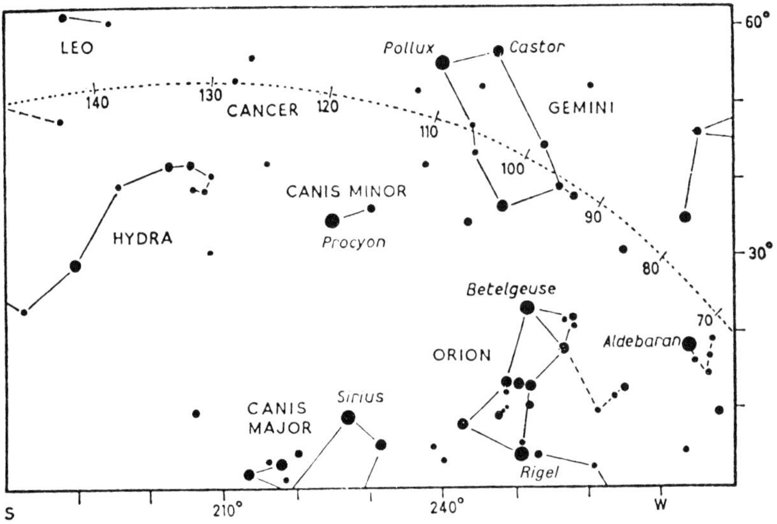

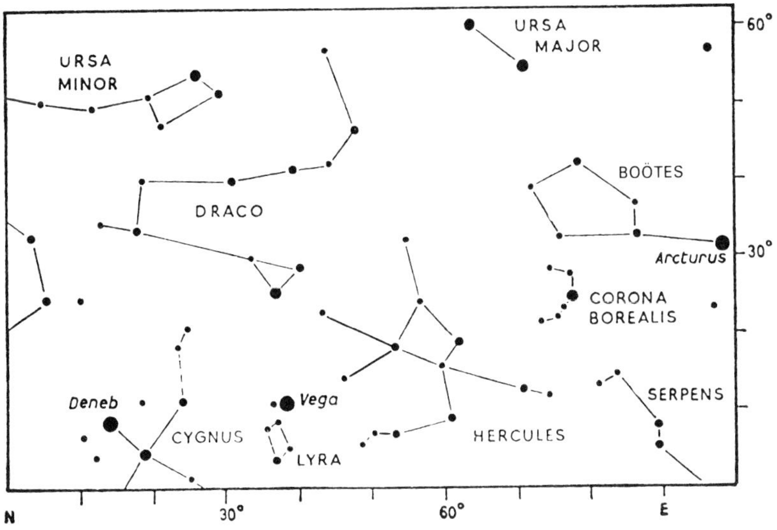

4L

January 6 at 5h	January 21 at 4h
February 6 at 3h	February 21 at 2h
March 6 at 1h	March 21 at midnight
April 6 at 23h	April 21 at 22h
May 6 at 21h	May 21 at 20h

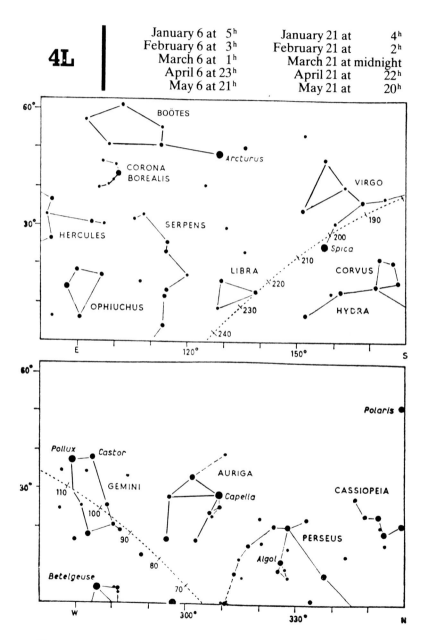

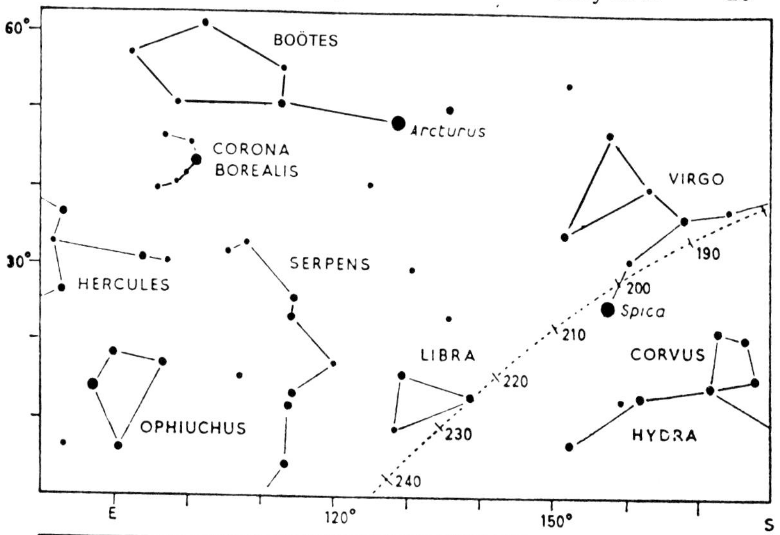

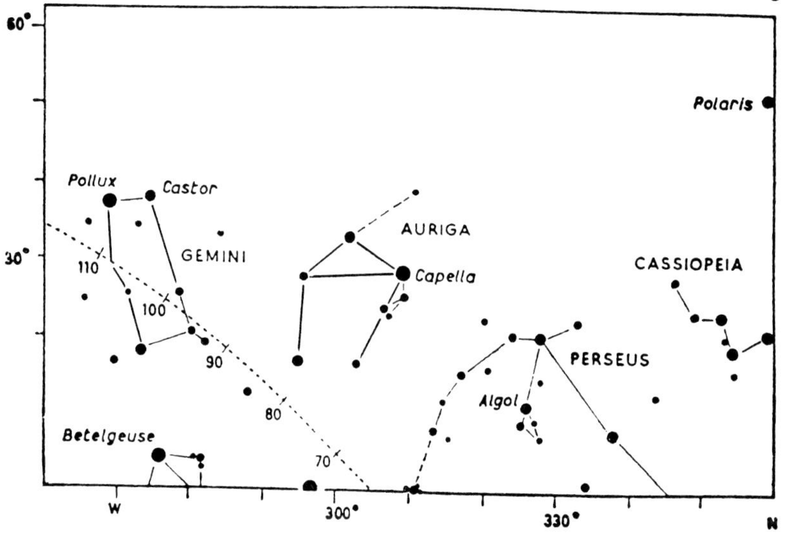

January 6 at 5h	January 21 at 4h
February 6 at 3h	February 21 at 2h
March 6 at 1h	March 21 at midnight
April 6 at 23h	April 21 at 22h
May 6 at 21h	May 21 at 20h

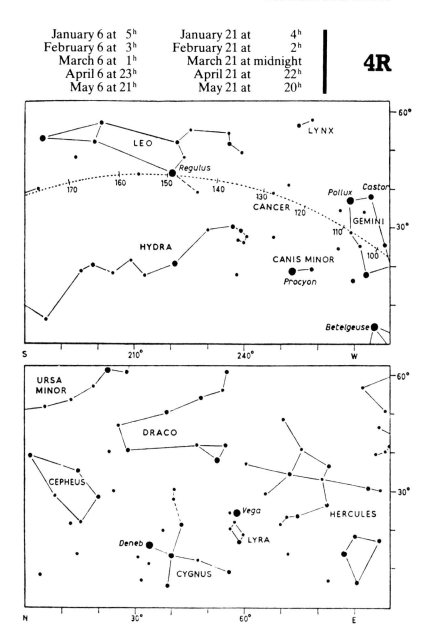

4R

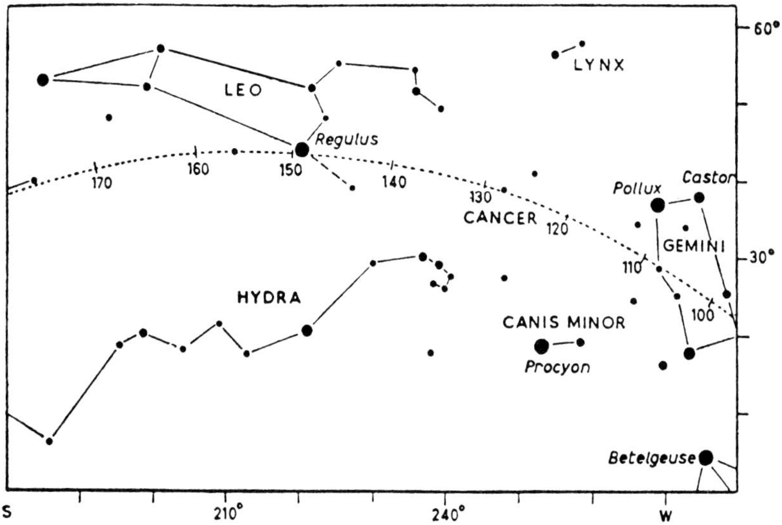

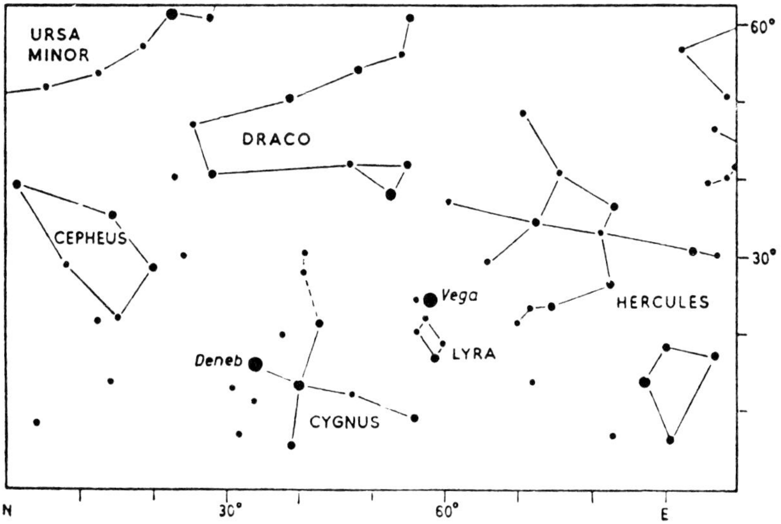

5L

January 6 at 7h	January 21 at 6h
February 6 at 5h	February 21 at 4h
March 6 at 3h	March 21 at 2h
April 6 at 1h	April 21 at midnight
May 6 at 23h	May 21 at 22h

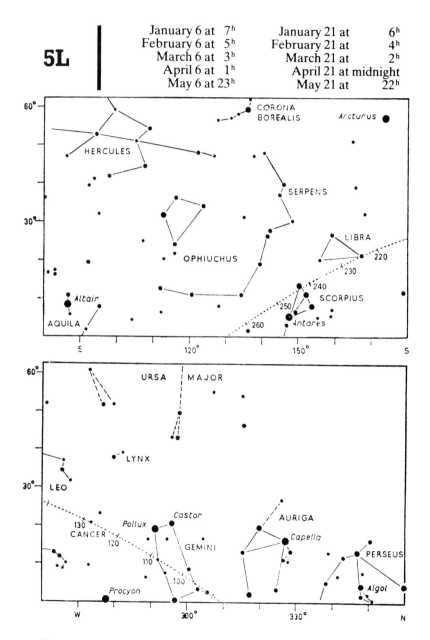

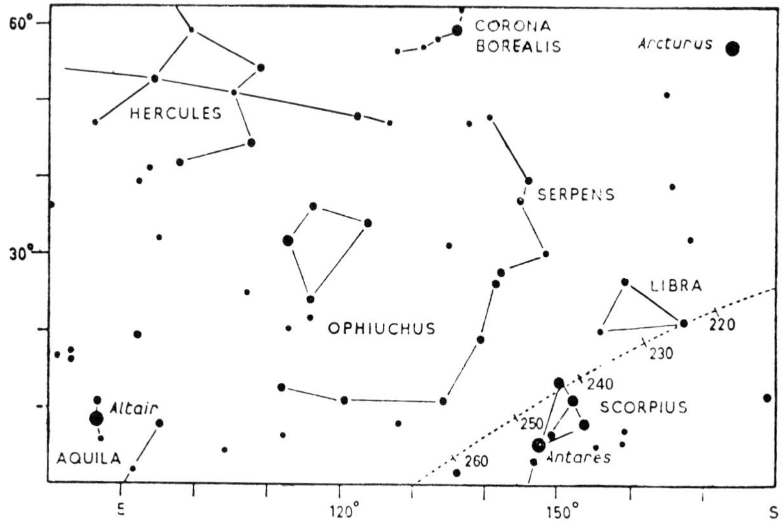

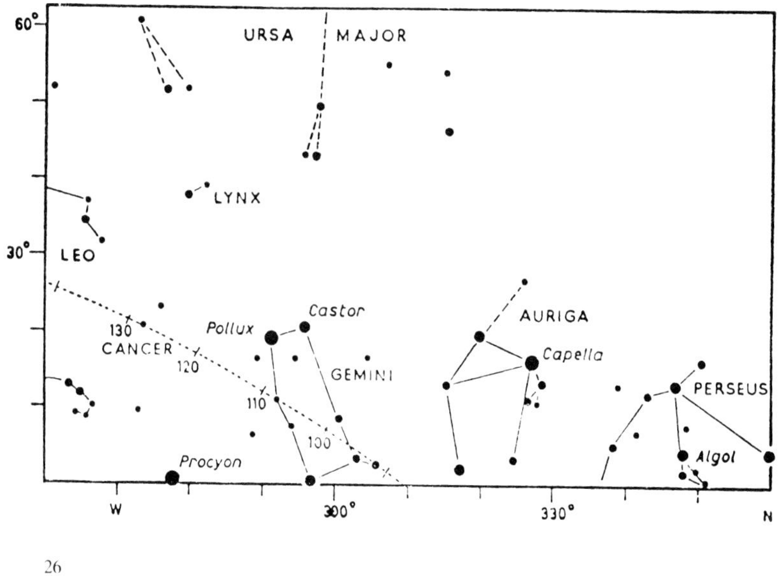

January 6 at 7h	January 21 at 6h	**5R**
February 6 at 5h	February 21 at 4h	
March 6 at 3h	March 21 at 2h	
April 6 at 1h	April 21 at midnight	
May 6 at 23h	May 21 at 22h	

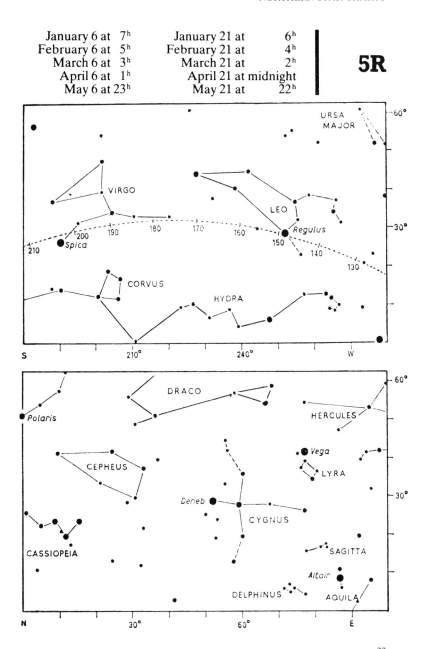

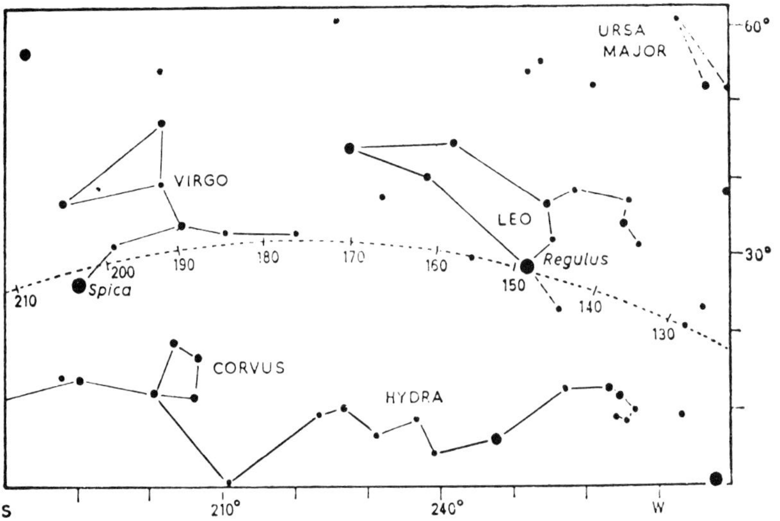

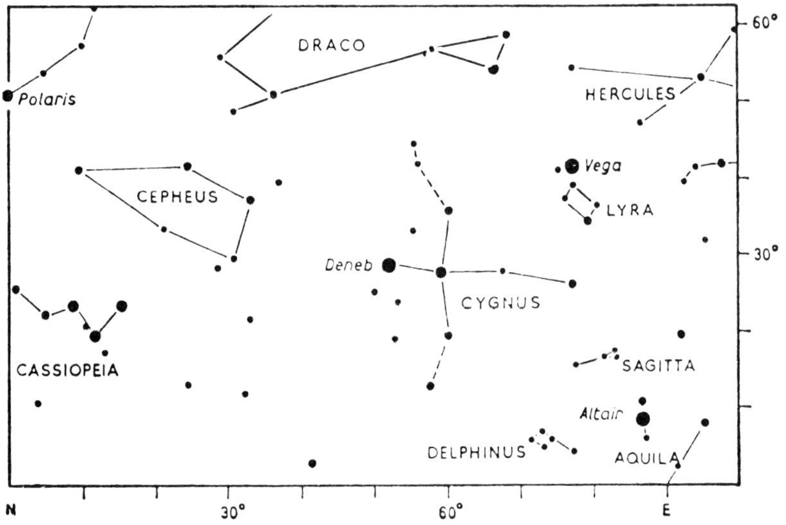

6L

March 6 at	5h	March 21 at	4h
April 6 at	3h	April 21 at	2h
May 6 at	1h	May 21 at	midnight
June 6 at	23h	June 21 at	22h
July 6 at	21h	July 21 at	20h

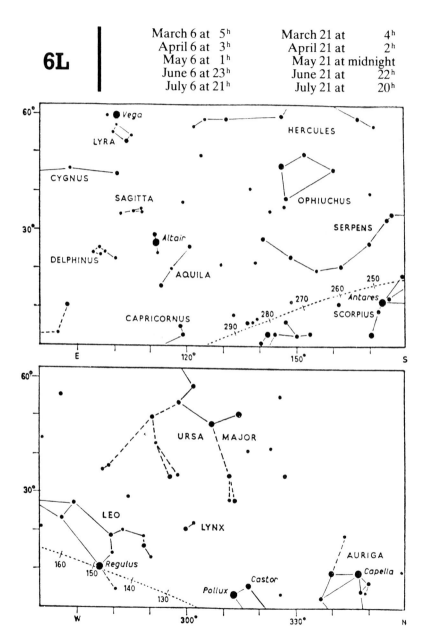

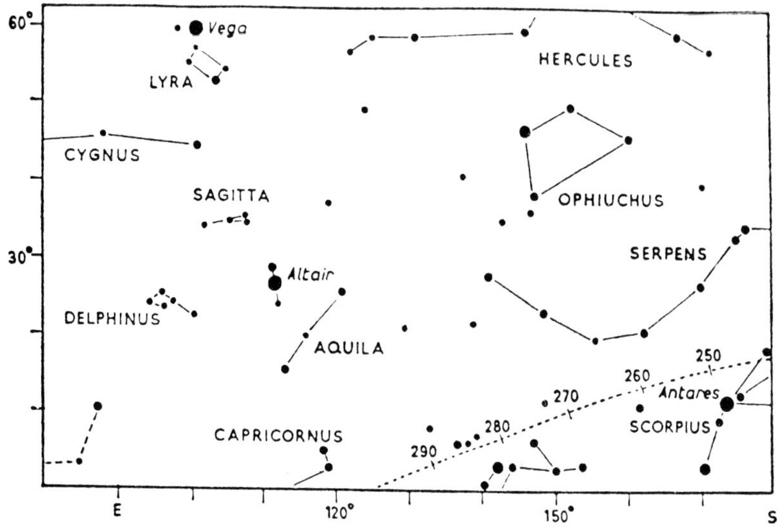

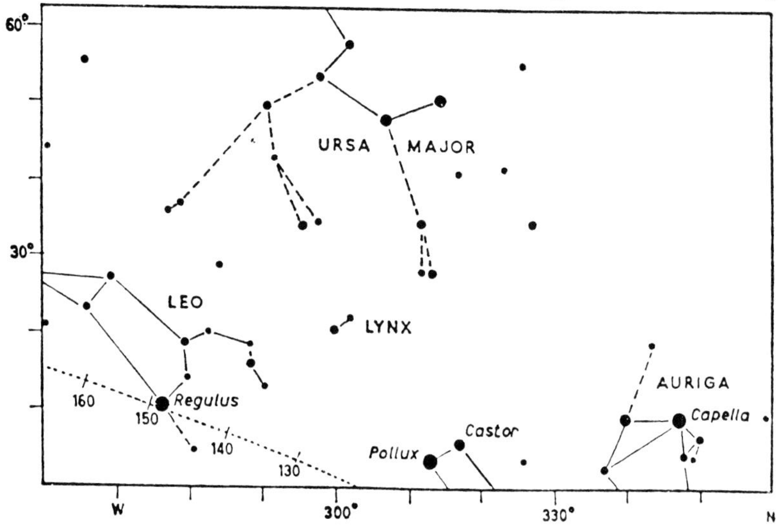

March 6 at 5h	March 21 at 4h
April 6 at 3h	April 21 at 2h
May 6 at 1h	May 21 at midnight
June 6 at 23h	June 21 at 22h
July 6 at 21h	July 21 at 20h

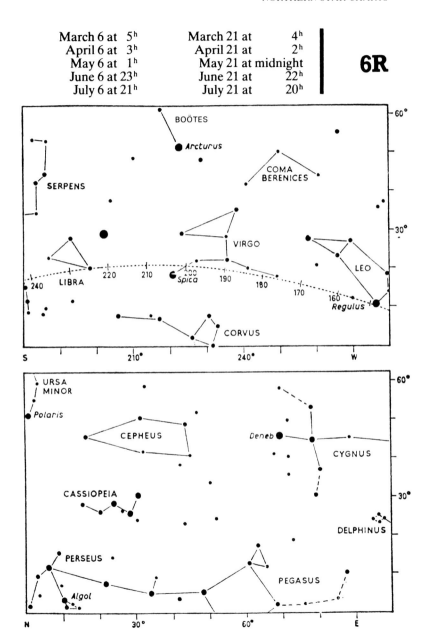

6R

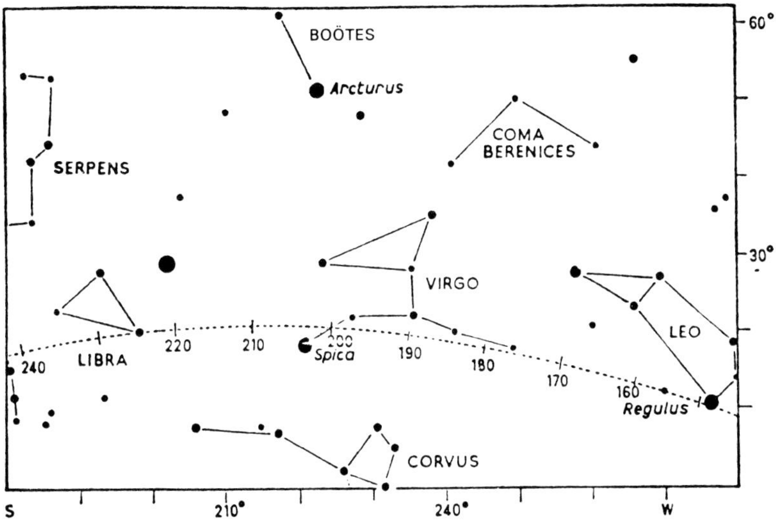

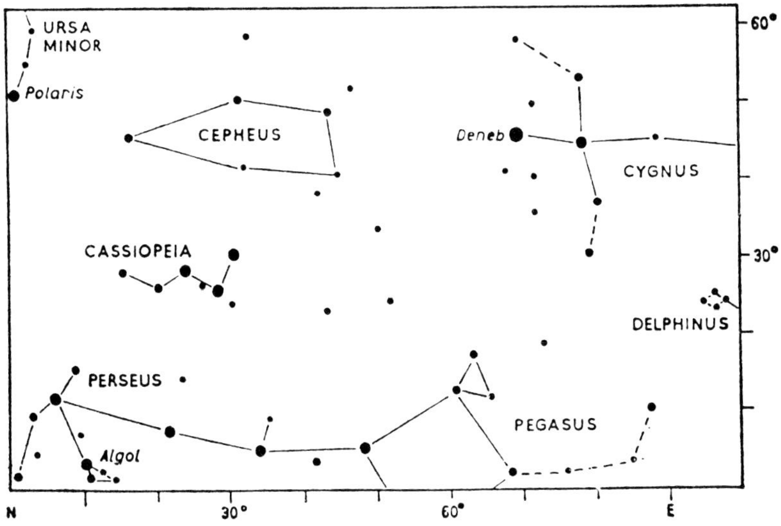

7L

May 6 at 3h	May 21 at 2h
June 6 at 1h	June 21 at midnight
July 6 at 23h	July 21 at 22h
August 6 at 21h	August 21 at 20h
September 6 at 19h	September 21 at 18h

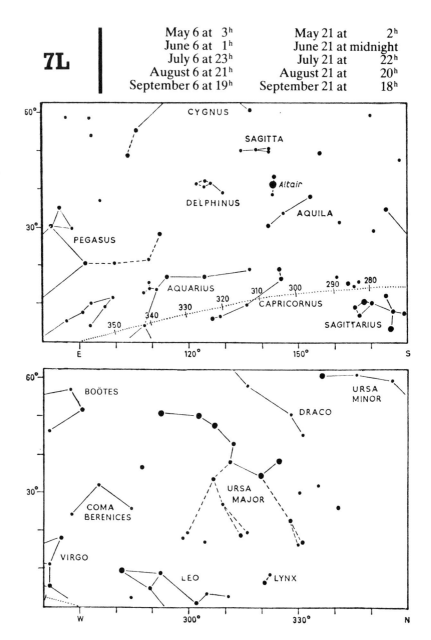

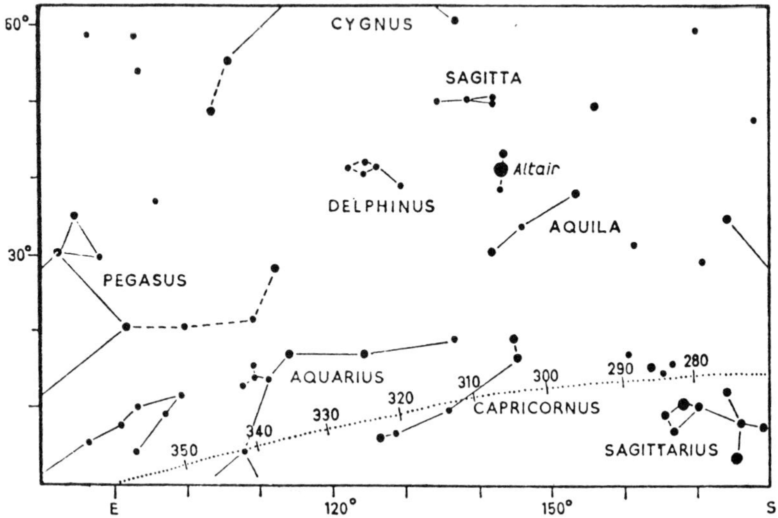

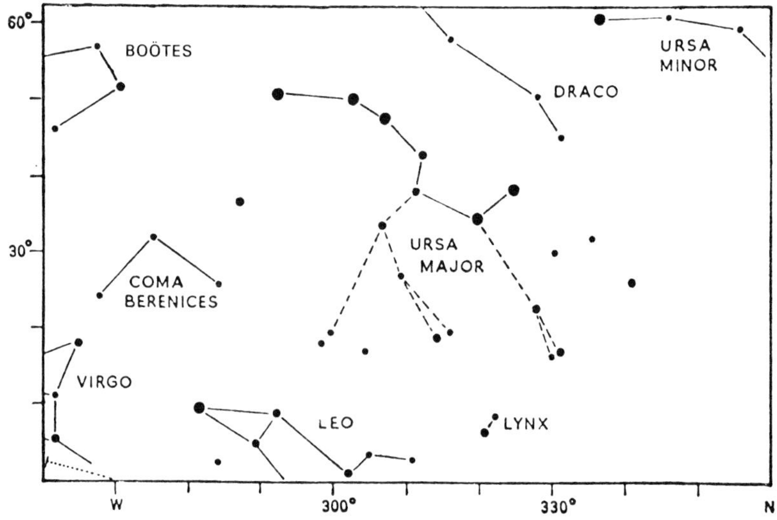

May 6 at	3^h	May 21 at	2^h
June 6 at	1^h	June 21 at	midnight
July 6 at	23^h	July 21 at	22^h
August 6 at	21^h	August 21 at	20^h
September 6 at	19^h	September 21 at	18^h

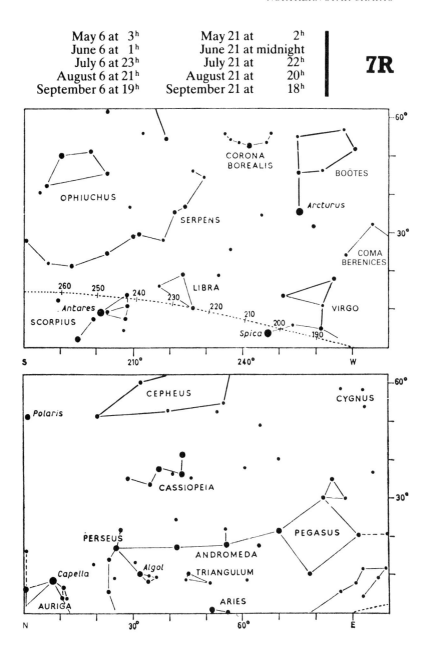

7R

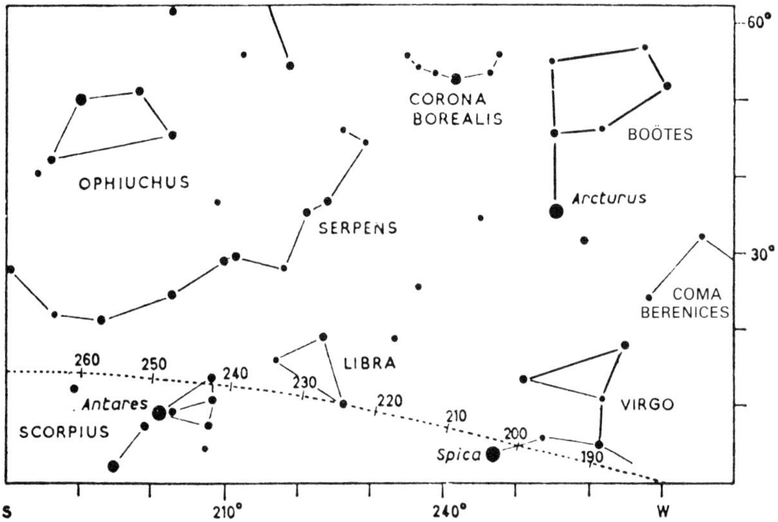

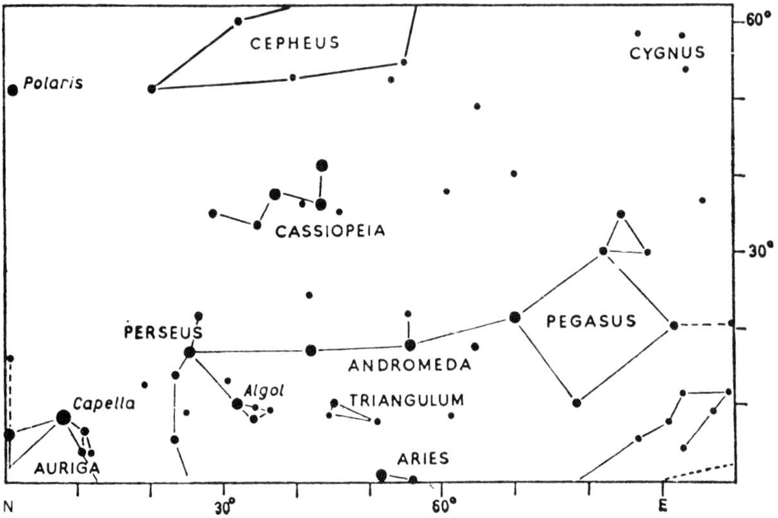

8L

July 6 at 1^h	July 21 at midnight
August 6 at 23^h	August 21 at 22^h
September 6 at 21^h	September 21 at 20^h
October 6 at 19^h	October 21 at 18^h
November 6 at 17^h	November 21 at 16^h

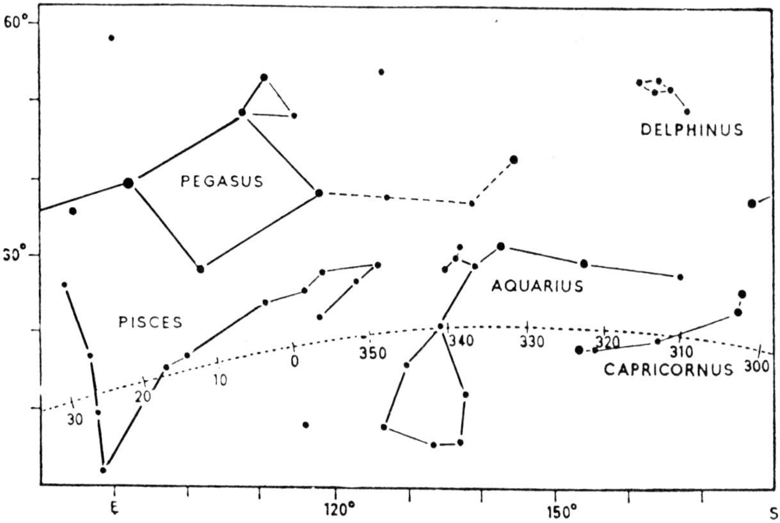

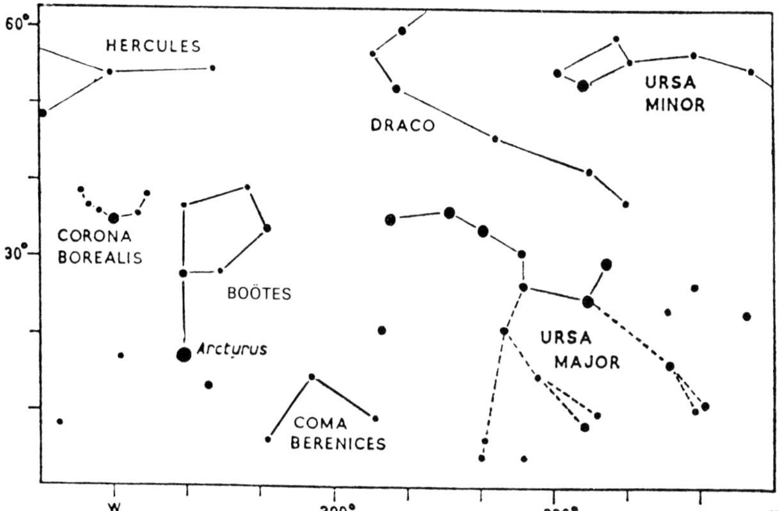

July 6 at 1^h	July 21 at midnight	**8R**
August 6 at 23^h	August 21 at 22^h	
September 6 at 21^h	September 21 at 20^h	
October 6 at 19^h	October 21 at 18^h	
November 6 at 17^h	November 21 at 16^h	

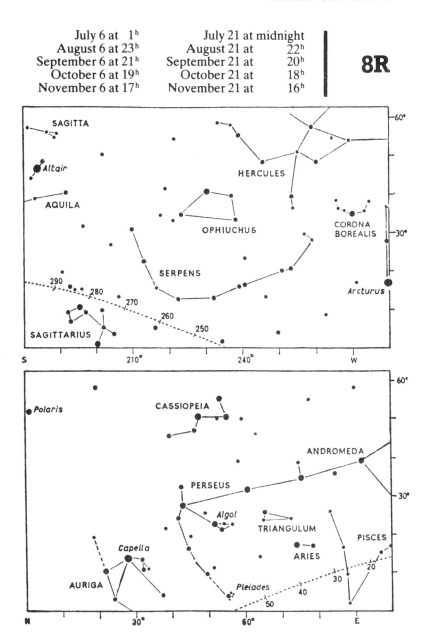

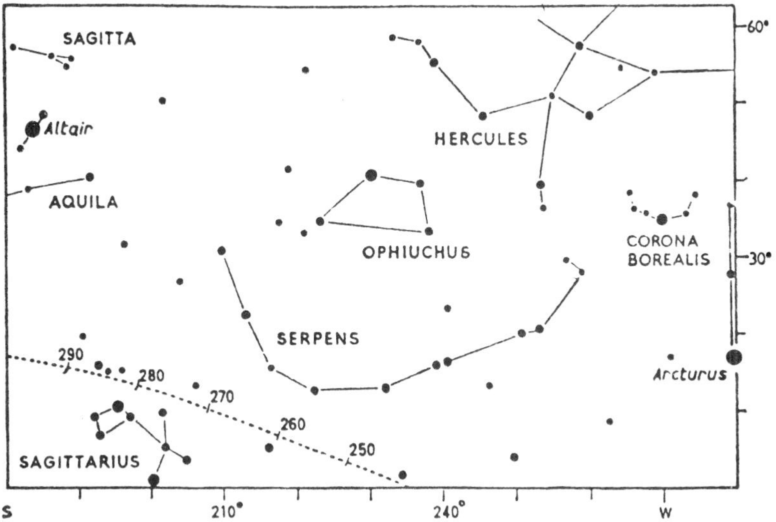

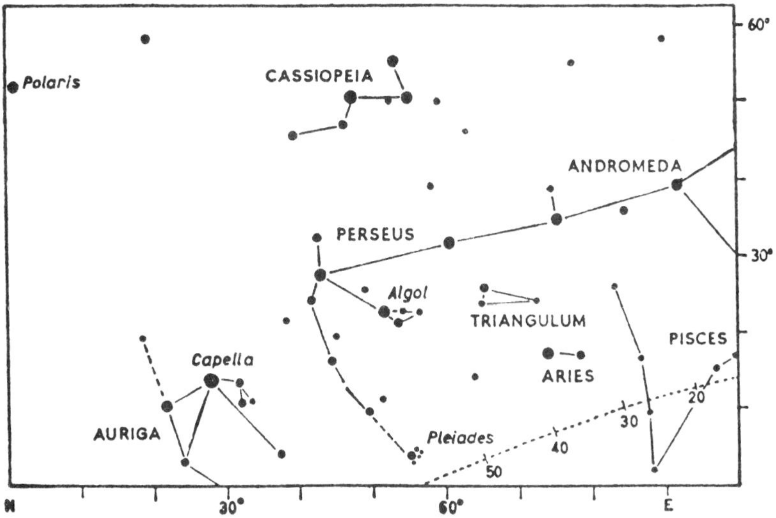

9L

August 6 at 1h	August 21 at midnight
September 6 at 23h	September 21 at 22h
October 6 at 21h	October 21 at 20h
November 6 at 19h	November 21 at 18h
December 6 at 17h	December 21 at 16h

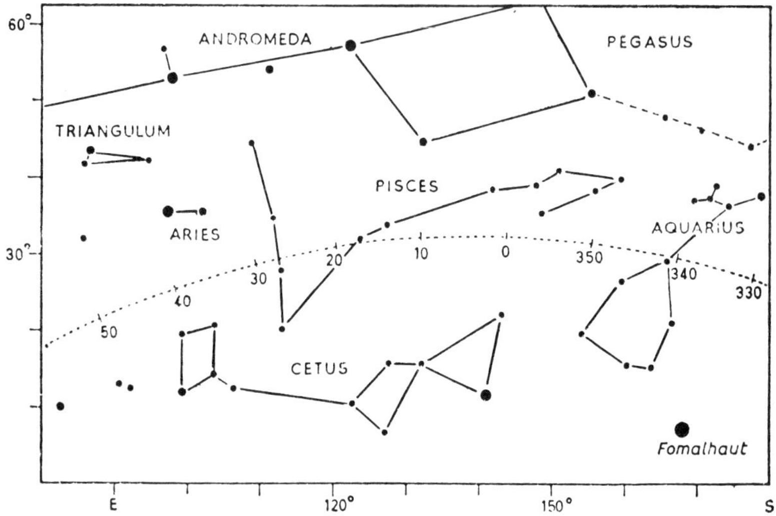

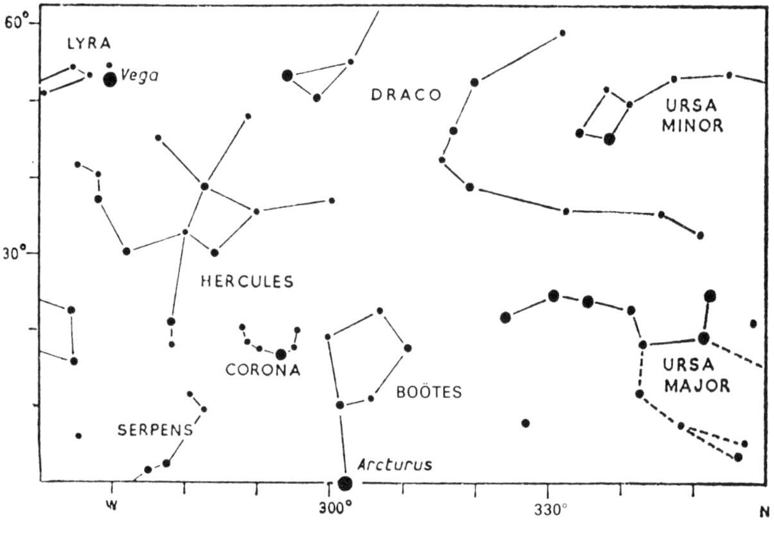

August 6 at 1ʰ	August 21 at midnight	**9R**
September 6 at 23ʰ	September 21 at 22ʰ	
October 6 at 21ʰ	October 21 at 20ʰ	
November 6 at 19ʰ	November 21 at 18ʰ	
December 6 at 17ʰ	December 21 at 16ʰ	

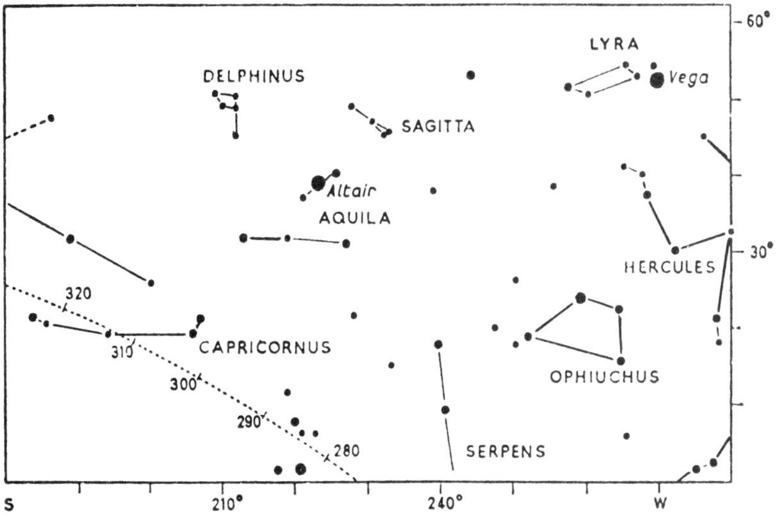

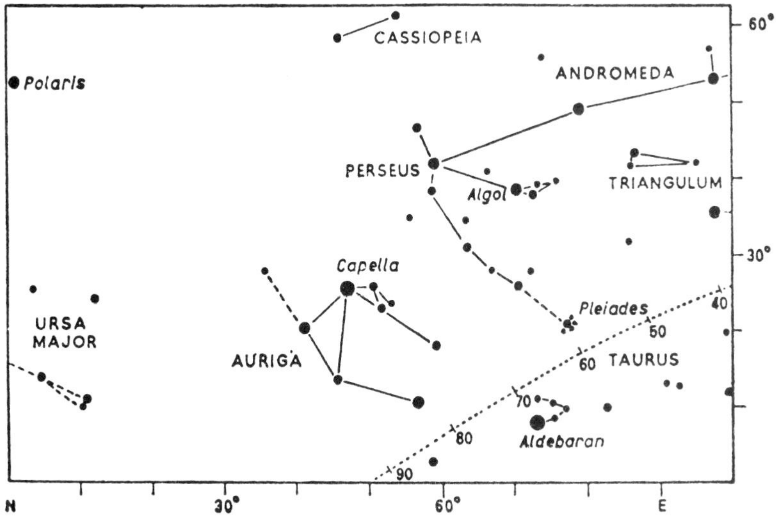

10L

August 6 at 3^h	August 21 at 2^h
September 6 at 1^h	September 21 at midnight
October 6 at 23^h	October 21 at 22^h
November 6 at 21^h	November 21 at 20^h
December 6 at 19^h	December 21 at 18^h

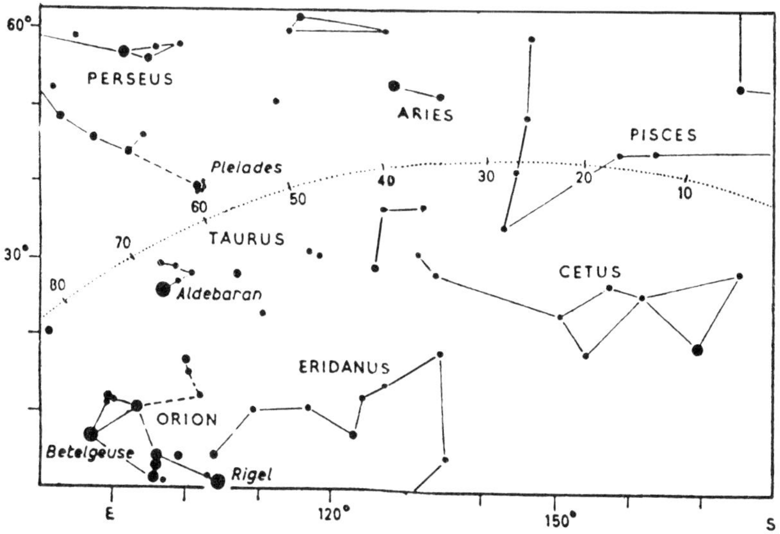

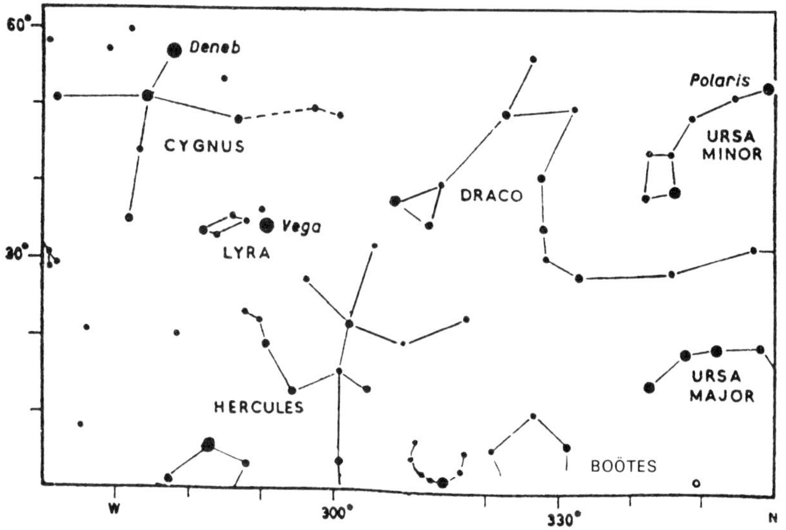

August 6 at 3h	August 21 at 2h
September 6 at 1h	September 21 at midnight
October 6 at 23h	October 21 at 22h
November 6 at 21h	November 21 at 20h
December 6 at 19h	December 21 at 18h

10R

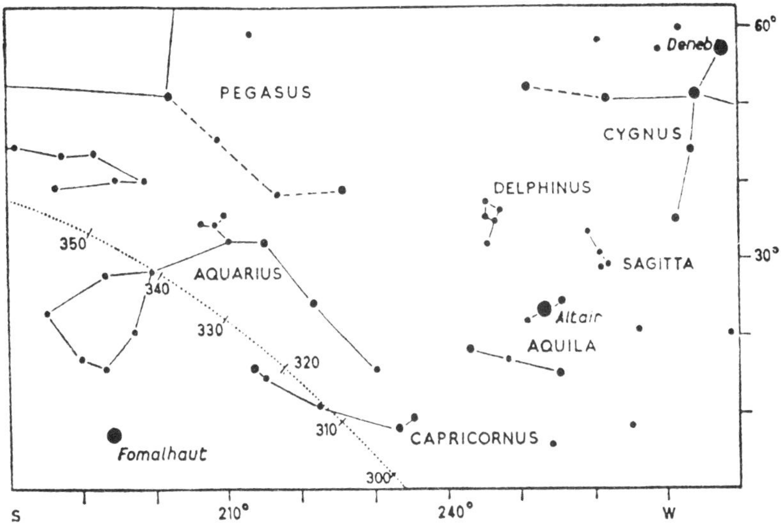

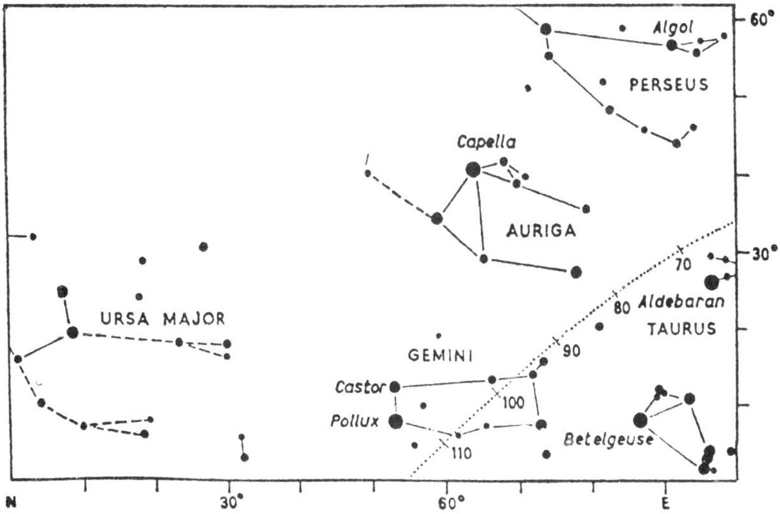

11L

September 6 at 3h	September 21 at 2h
October 6 at 1h	October 21 at midnight
November 6 at 23h	November 21 at 22h
December 6 at 21h	December 21 at 20h
January 6 at 19h	January 21 at 18h

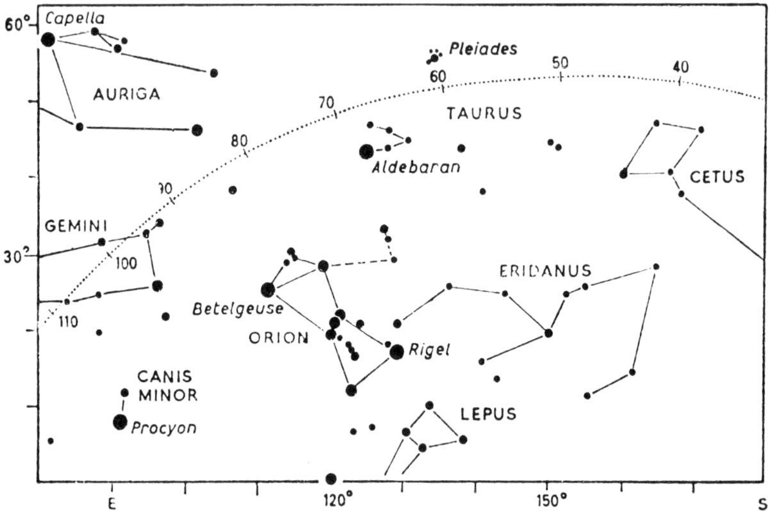

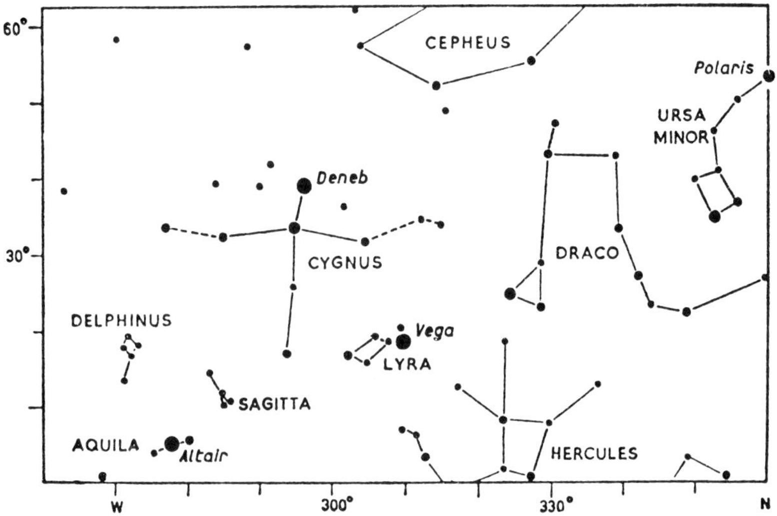

September 6 at 3h	September 21 at 2h
October 6 at 1h	October 21 at midnight
November 6 at 23h	November 21 at 22h
December 6 at 21h	December 21 at 20h
January 6 at 19h	January 21 at 18h

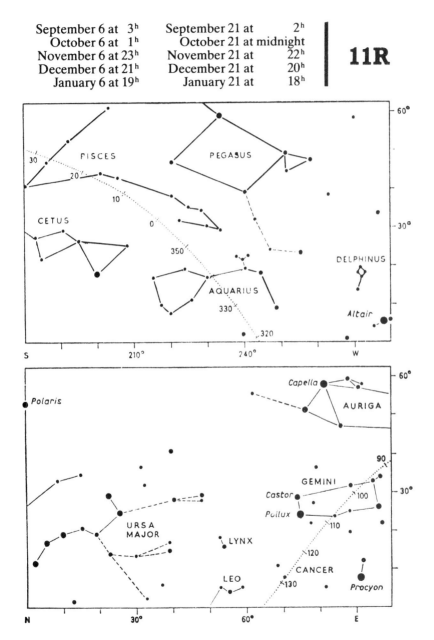

11R

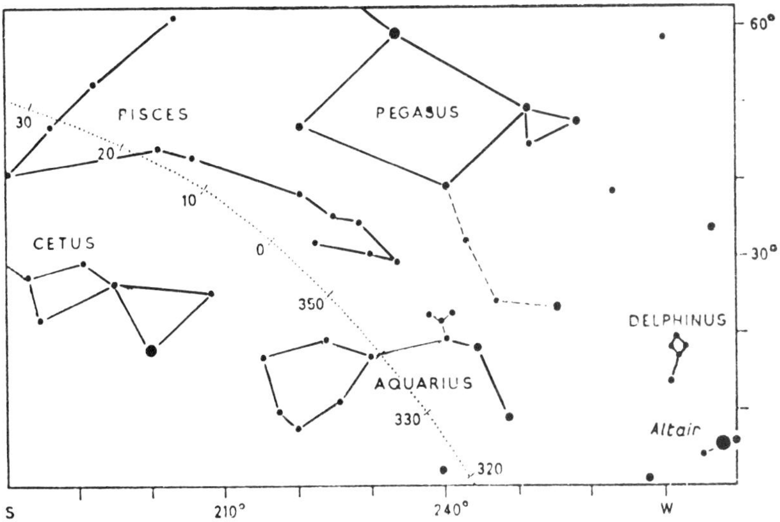

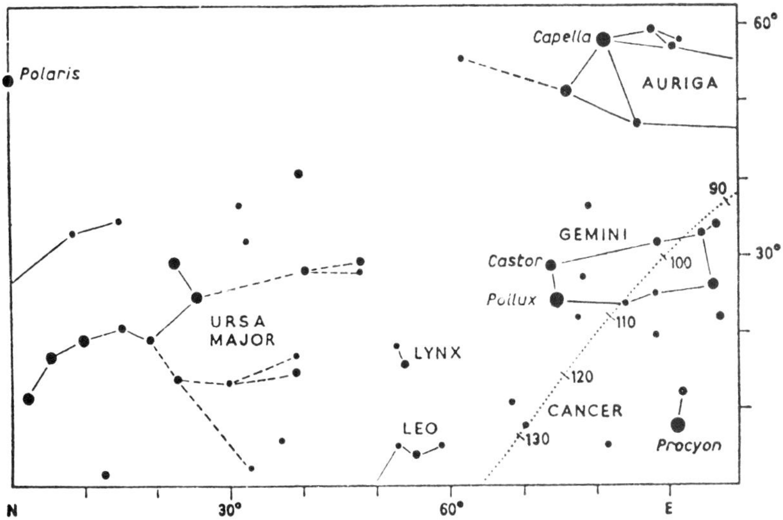

12L

October 6 at 3h	October 21 at 2h
November 6 at 1h	November 21 at midnight
December 6 at 23h	December 21 at 22h
January 6 at 21h	January 21 at 20h
February 6 at 19h	February 21 at 18h

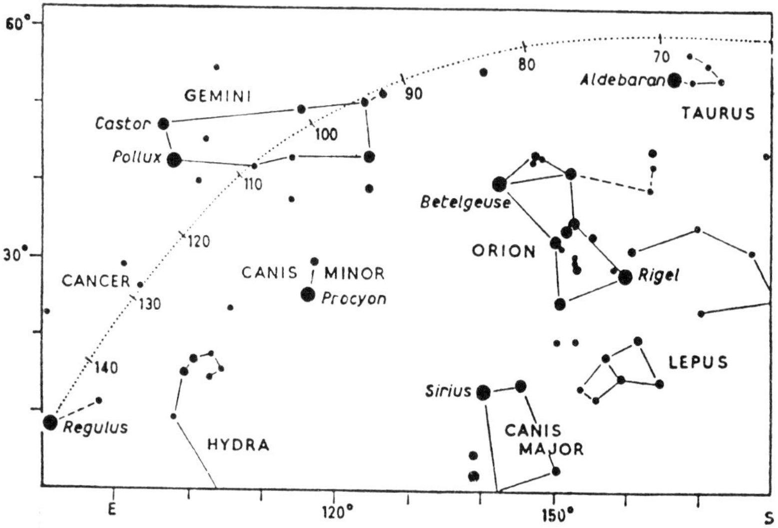

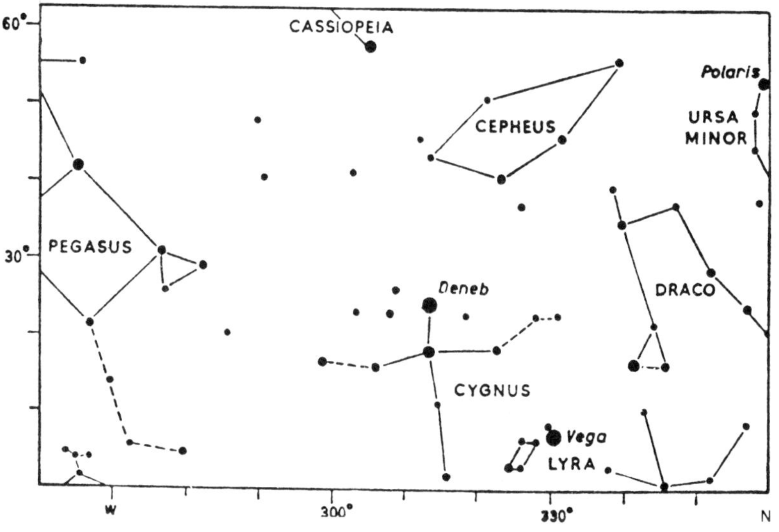

October 6 at 3h	October 21 at 2h	**12R**
November 6 at 1h	November 21 at midnight	
December 6 at 23h	December 21 at 22h	
January 6 at 21h	January 21 at 20h	
February 6 at 19h	February 21 at 18h	

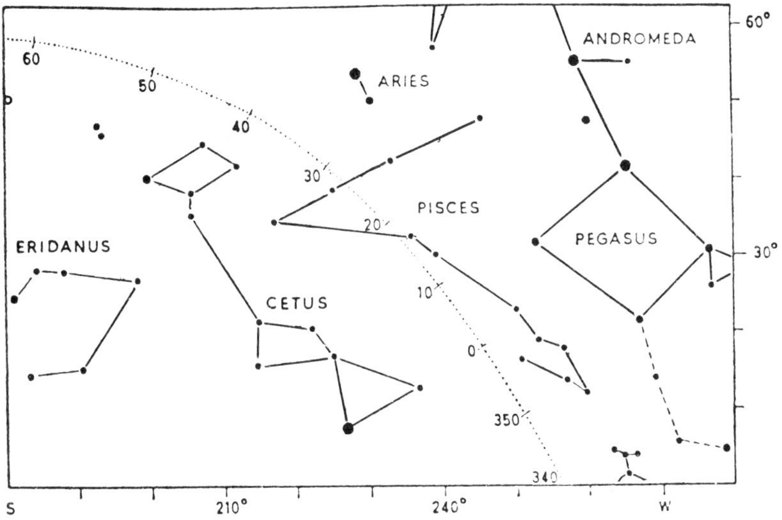

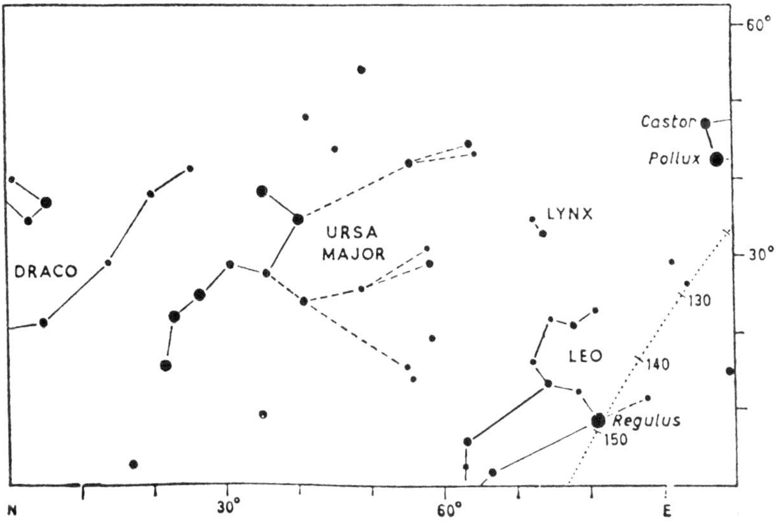

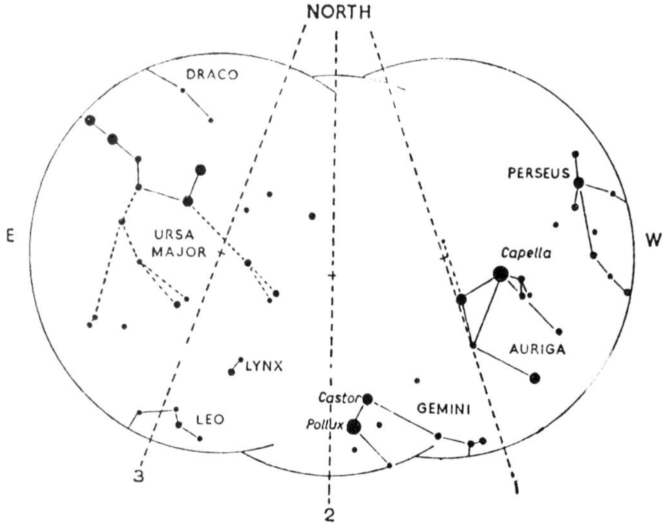

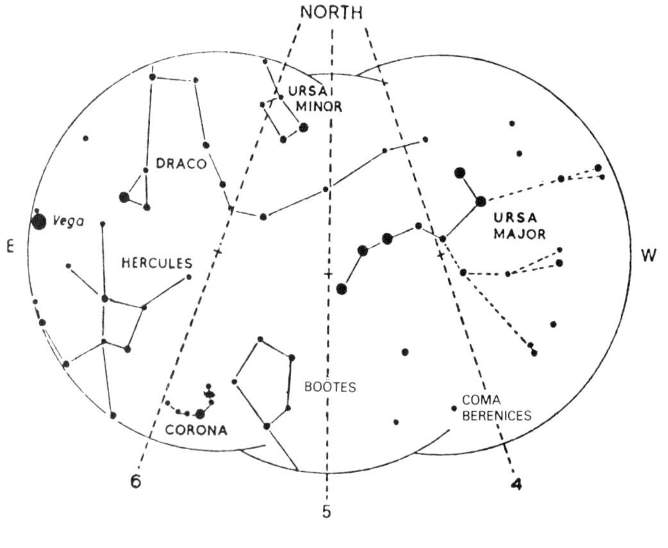

Northern Hemisphere Overhead Stars

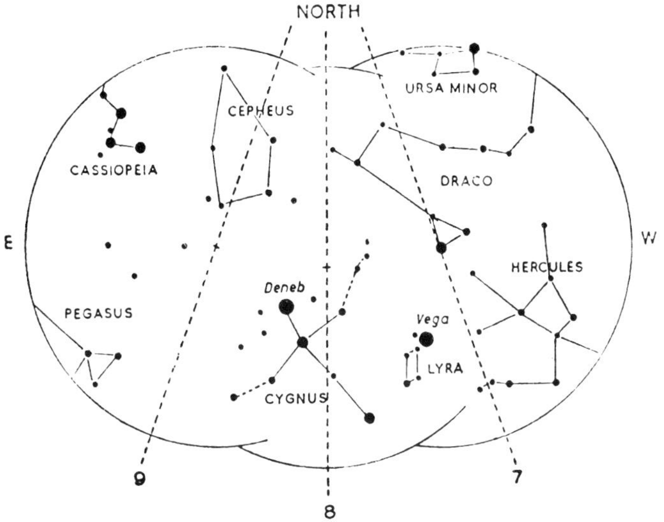

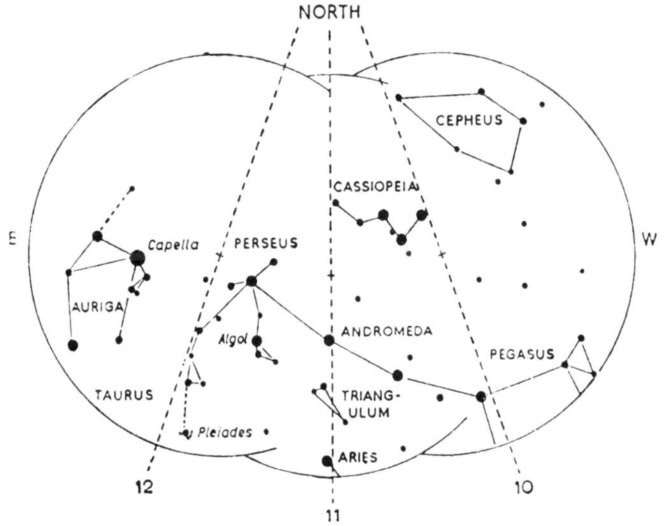

Northern Hemisphere Overhead Stars

1L

October 6 at 5^h	October 21 at 4^h
November 6 at 3^h	November 21 at 2^h
December 6 at 1^h	December 21 at midnight
January 6 at 23^h	January 21 at 22^h
February 6 at 21^h	February 21 at 20^h

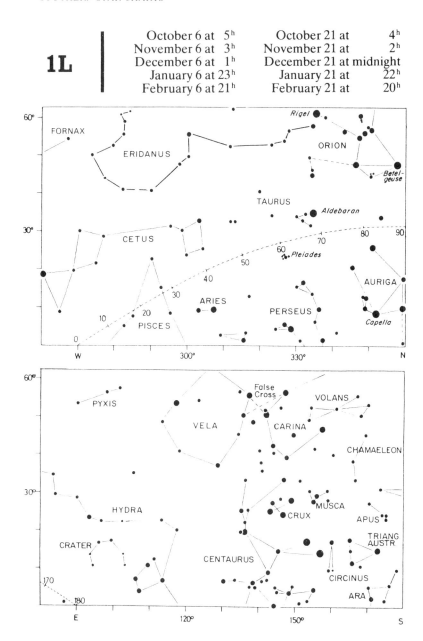

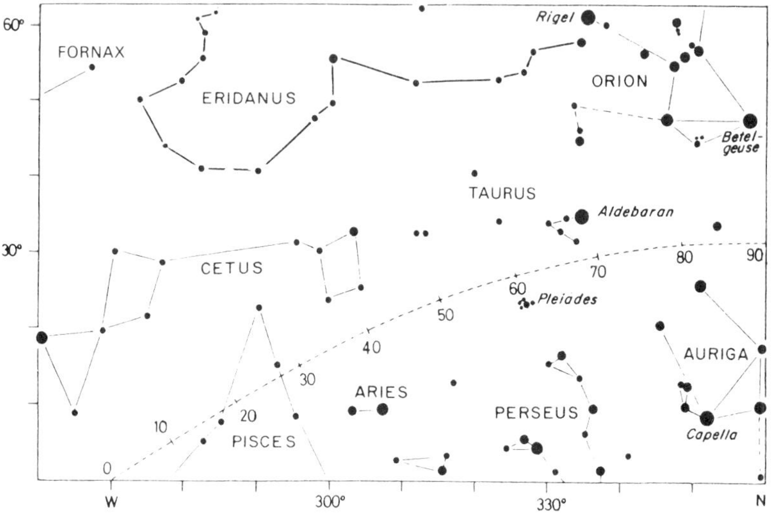

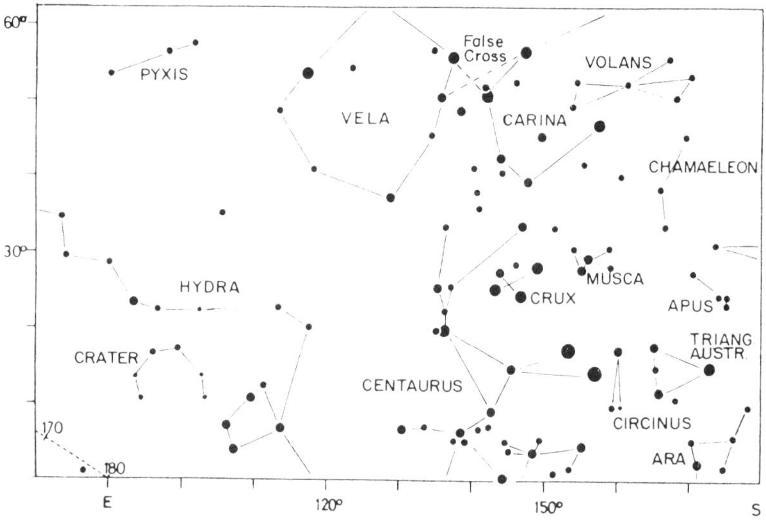

October 6 at 5h	October 21 at 4h	**1R**
November 6 at 3h	November 21 at 2h	
December 6 at 1h	December 21 at midnight	
January 6 at 23h	January 21 at 22h	
February 6 at 21h	February 21 at 20h	

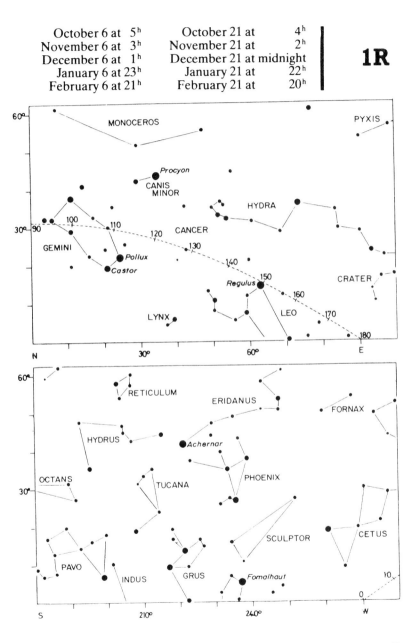

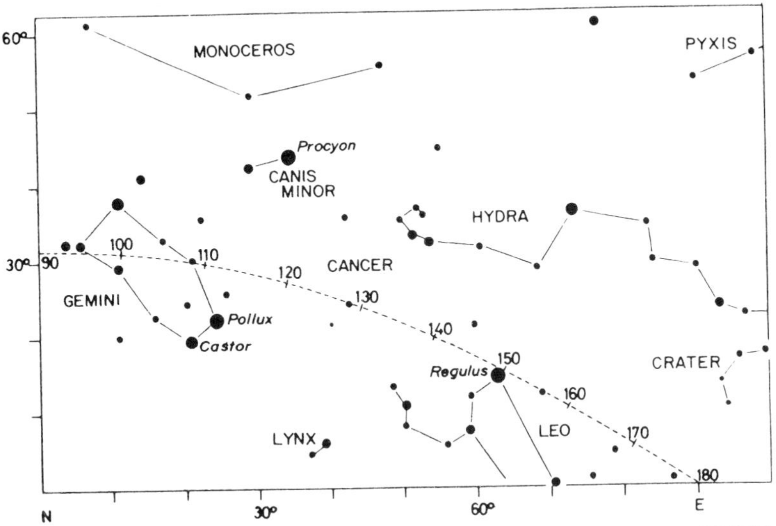

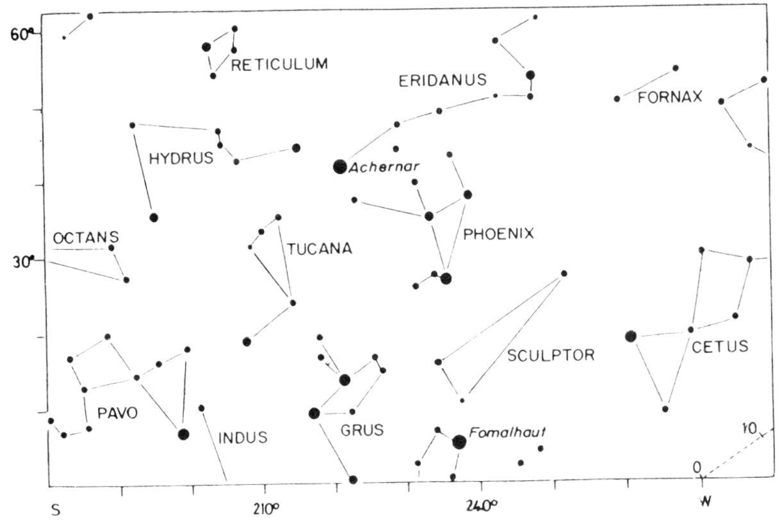

2L

November 6 at 5^h	November 21 at 4^h
December 6 at 3^h	December 21 at 2^h
January 6 at 1^h	January 21 at midnight
February 6 at 23^h	February 21 at 22^h
March 6 at 21^h	March 21 at 20^h

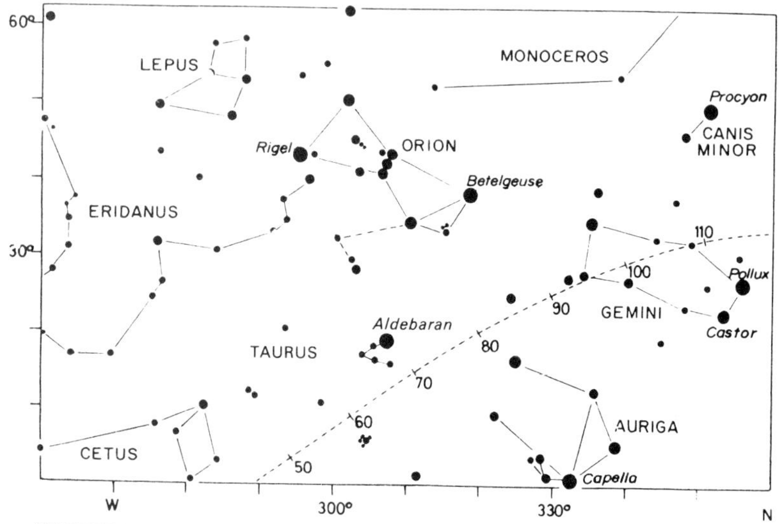

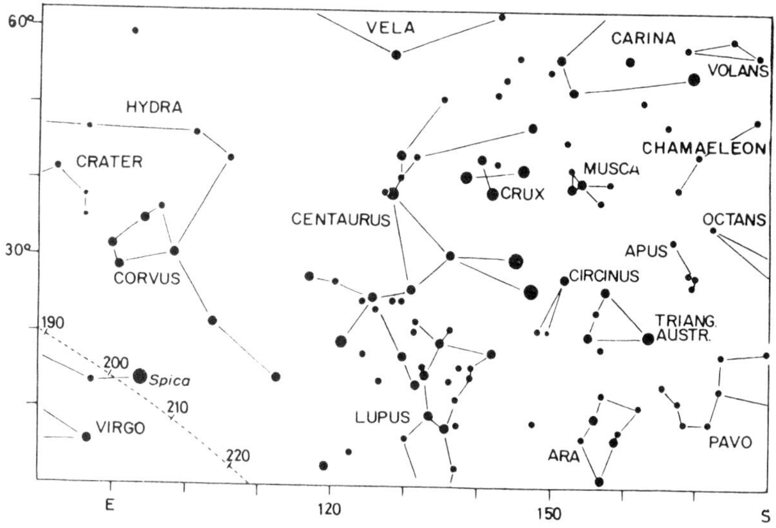

November 6 at 5ʰ	November 21 at 4ʰ	**2R**
December 6 at 3ʰ	December 21 at 2ʰ	
January 6 at 1ʰ	January 21 at midnight	
February 6 at 23ʰ	February 21 at 22ʰ	
March 6 at 21ʰ	March 21 at 20ʰ	

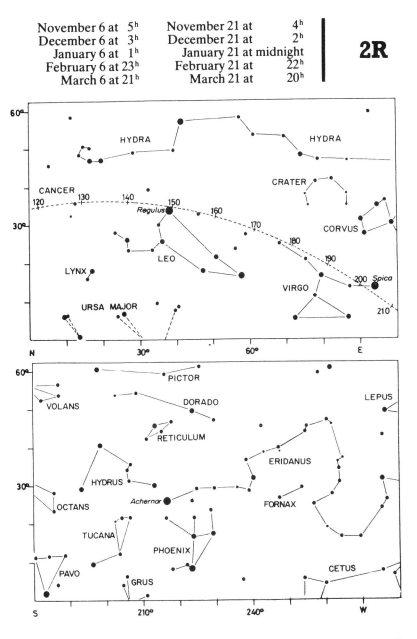

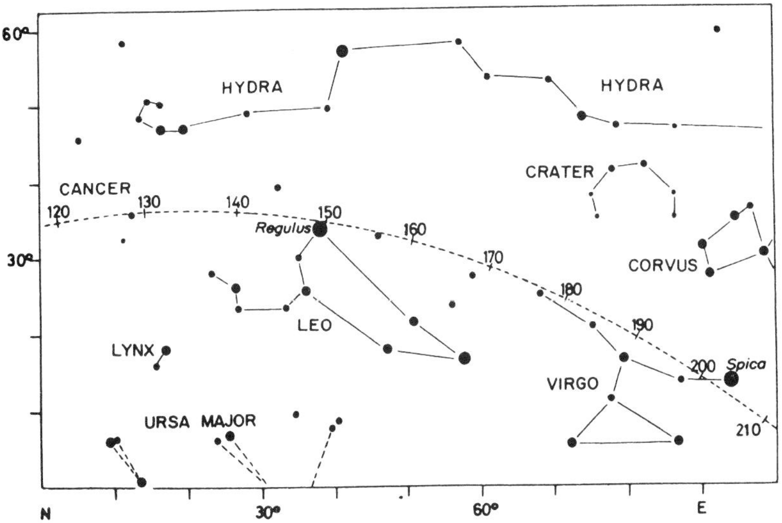

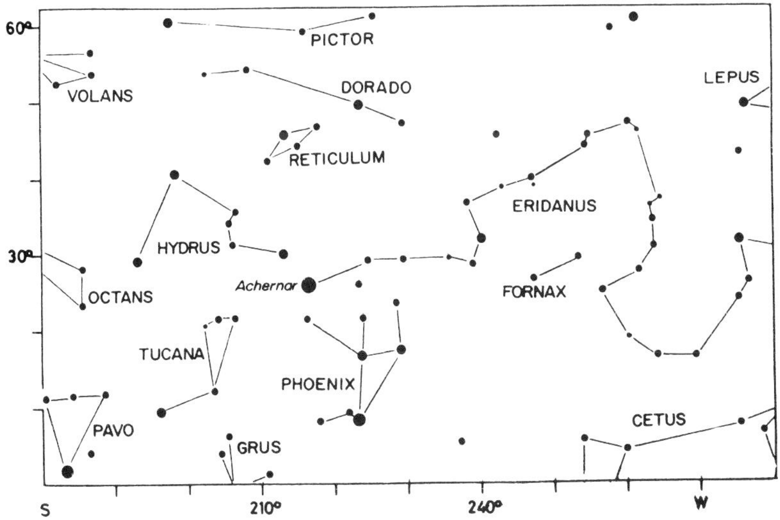

3L

January 6 at 3^h	January 21 at 2^h
February 6 at 1^h	February 21 at midnight
March 6 at 23^h	March 21 at 22^h
April 6 at 21^h	April 21 at 20^h
May 6 at 19^h	May 21 at 18^h

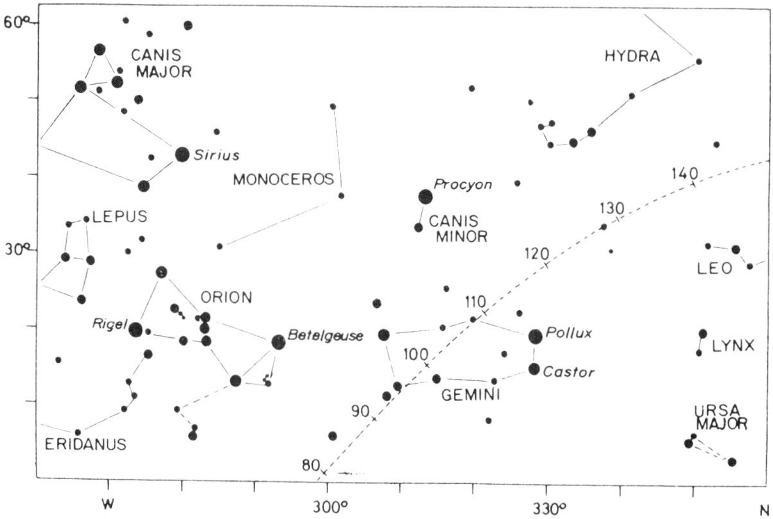

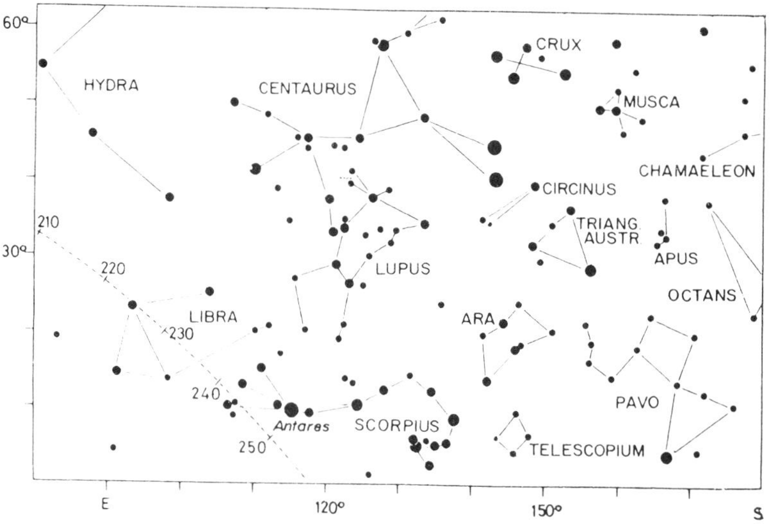

January 6 at 3h	January 21 at 2h
February 6 at 1h	February 21 at midnight
March 6 at 23h	March 21 at 22h
April 6 at 21h	April 21 at 20h
May 6 at 19h	May 21 at 18h

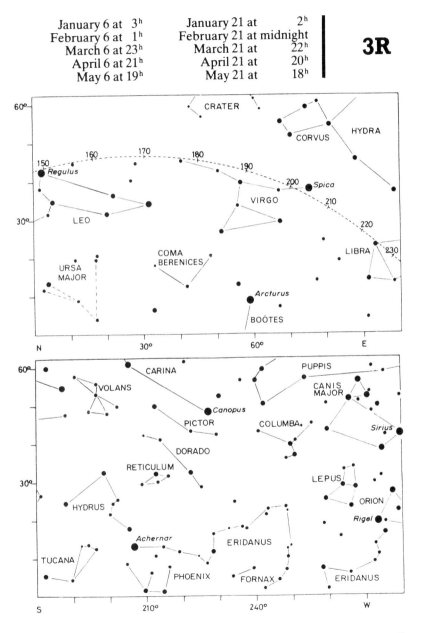

3R

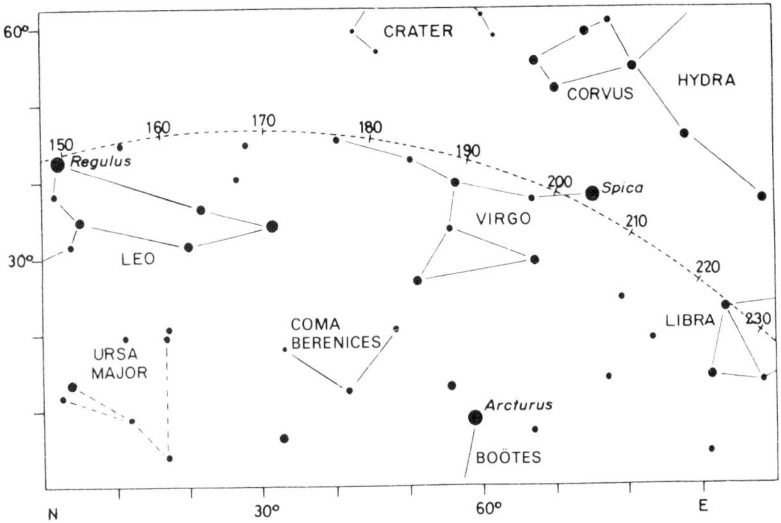

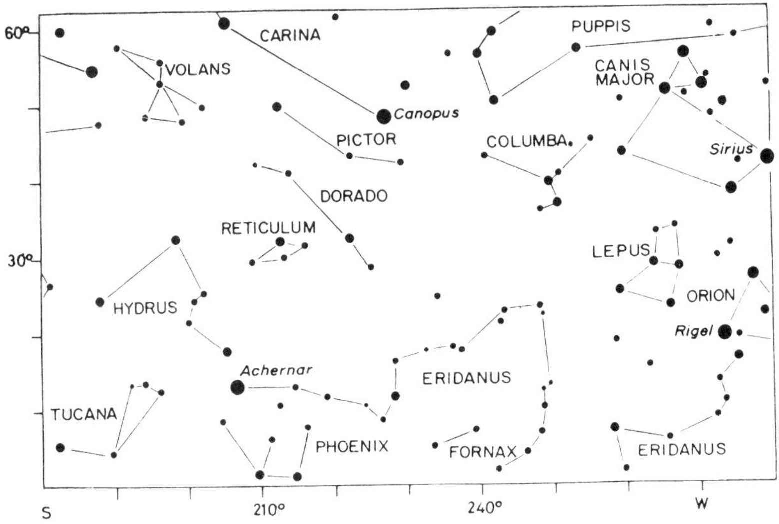

4L

February 6 at 3h	February 21 at 2h
March 6 at 1h	March 21 at midnight
April 6 at 23h	April 21 at 22h
May 6 at 21h	May 21 at 20h
June 6 at 19h	June 21 at 18h

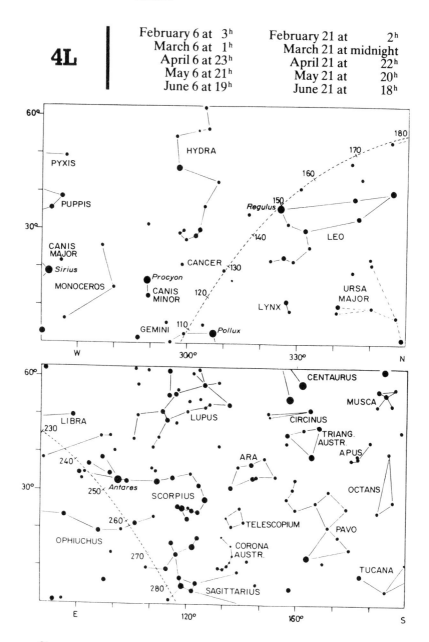

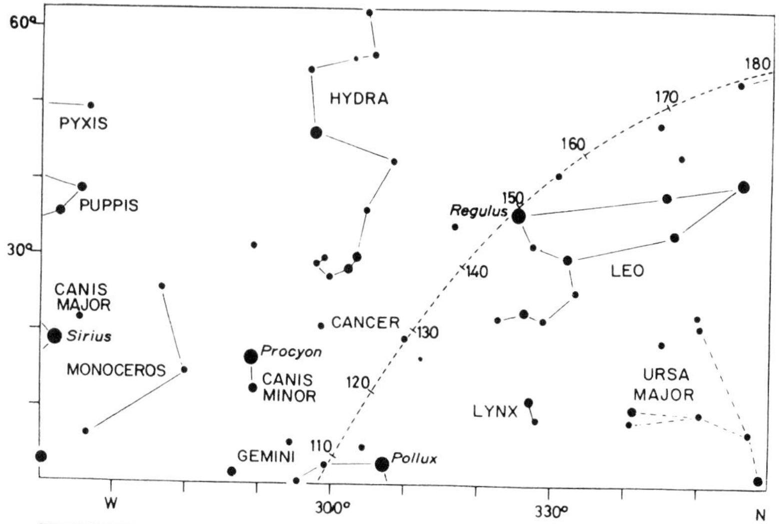

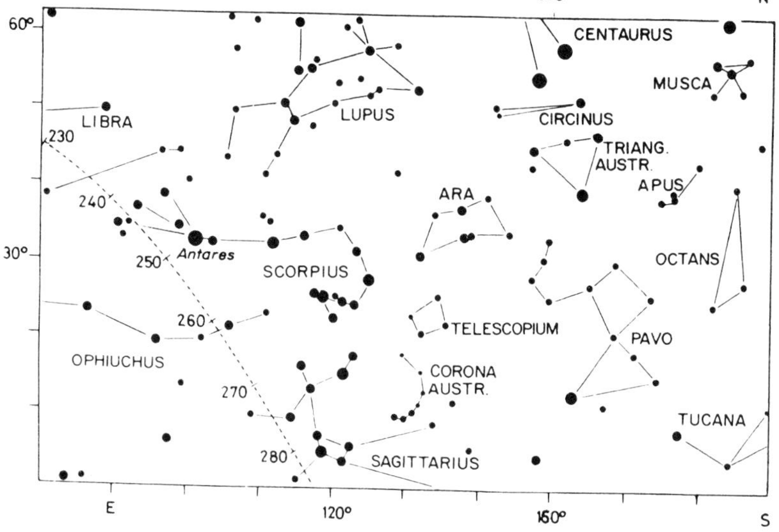

February 6 at 3ʰ	February 21 at 2ʰ	**4R**
March 6 at 1ʰ	March 21 at midnight	
April 6 at 23ʰ	April 21 at 22ʰ	
May 6 at 21ʰ	May 21 at 20ʰ	
June 6 at 19ʰ	June 21 at 18ʰ	

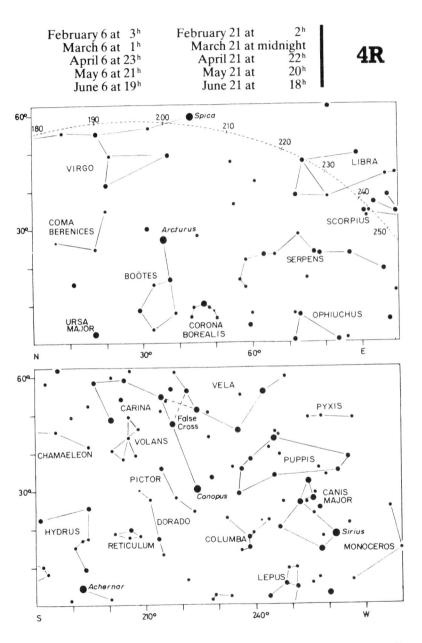

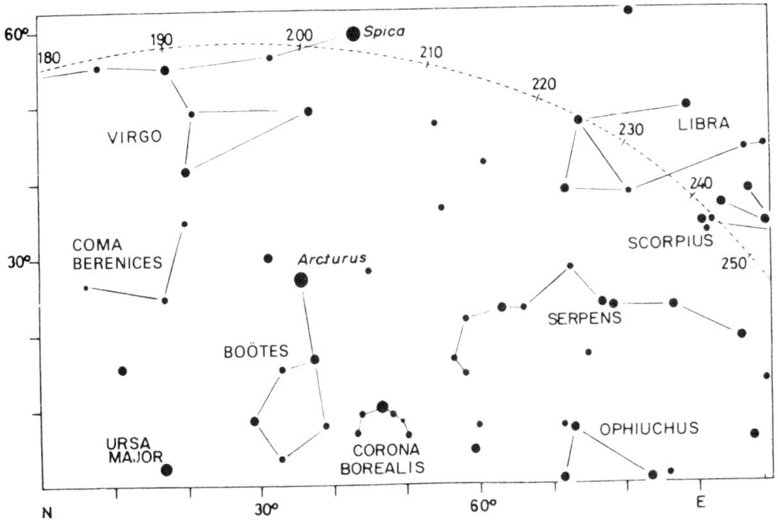

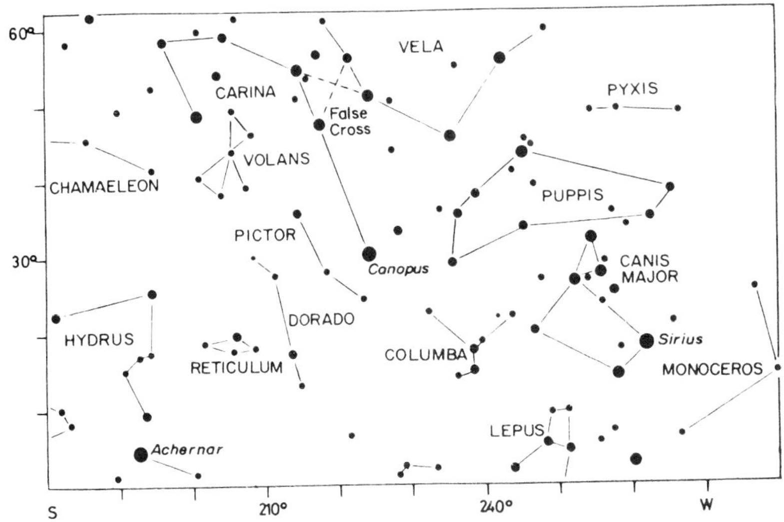

5L

March 6 at 3h	March 21 at 2h
April 6 at 1h	April 21 at midnight
May 6 at 23h	May 21 at 22h
June 6 at 21h	June 21 at 20h
July 6 at 19h	July 21 at 18h

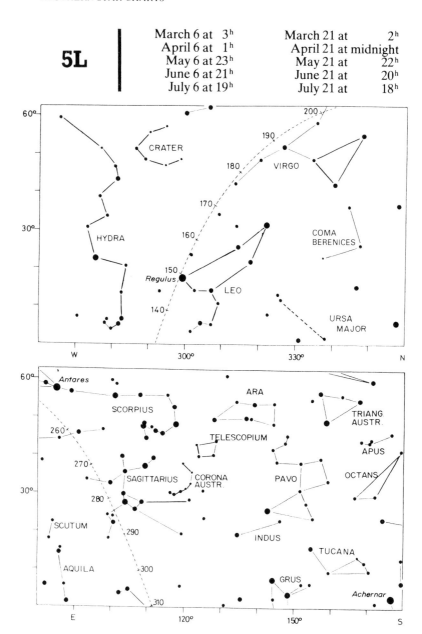

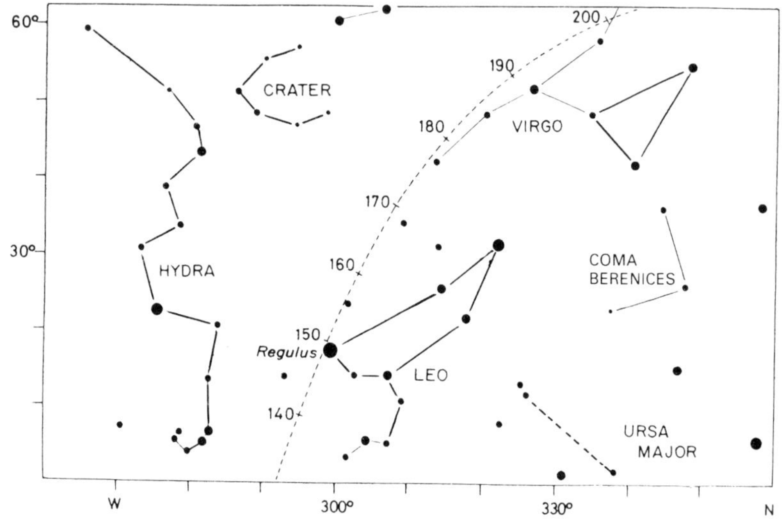

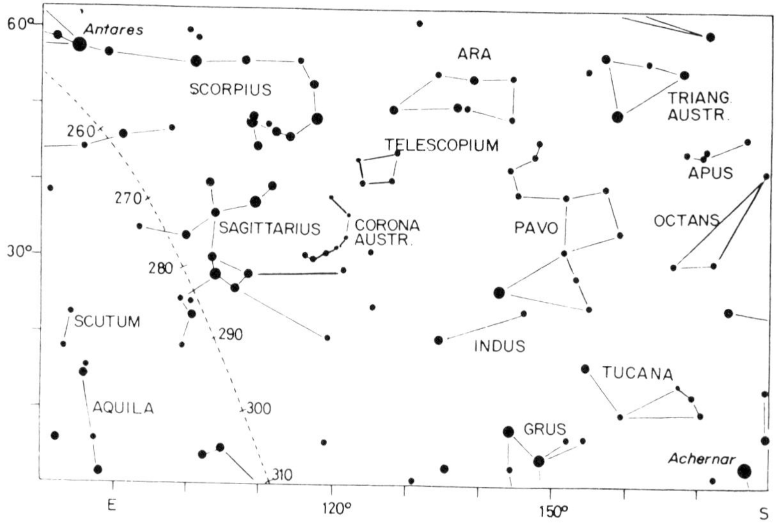

March 6 at 3h	March 21 at 2h	**5R**
April 6 at 1h	April 21 at midnight	
May 6 at 23h	May 21 at 22h	
June 6 at 21h	June 21 at 20h	
July 6 at 19h	July 21 at 18h	

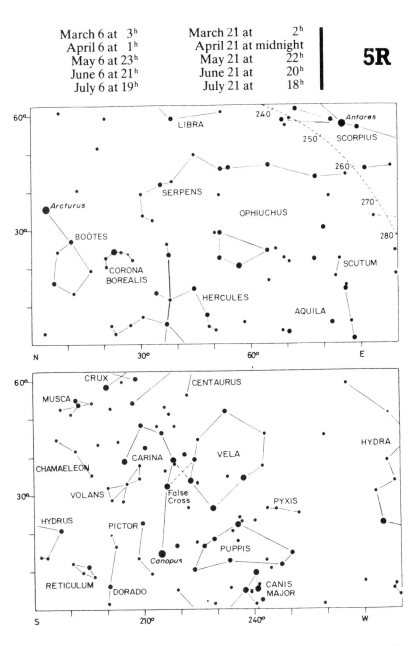

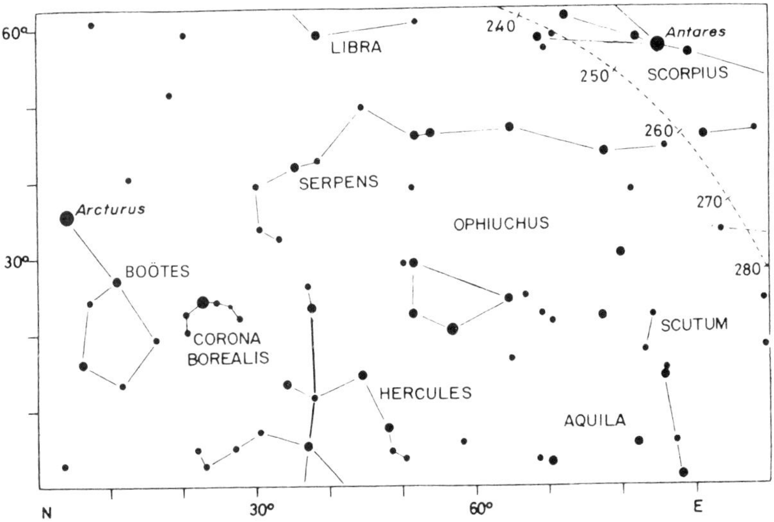

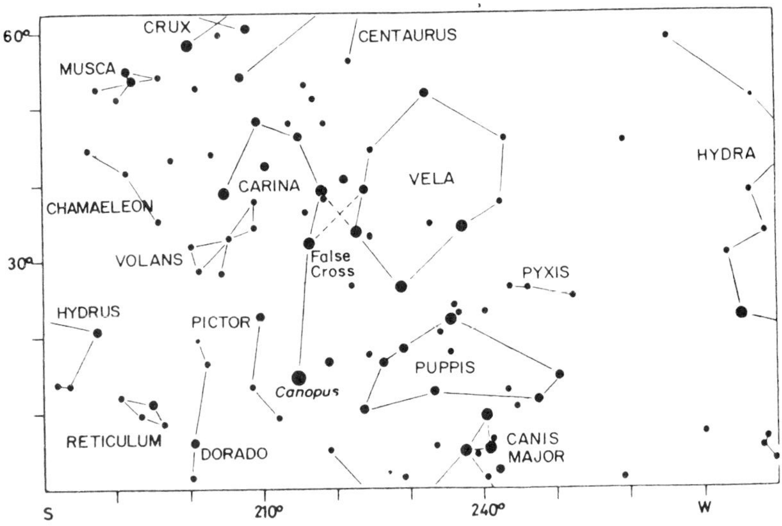

6L

March 6 at 5h	March 21 at 4h
April 6 at 3h	April 21 at 2h
May 6 at 1h	May 21 at midnight
June 6 at 23h	June 21 at 22h
July 6 at 21h	July 21 at 20h

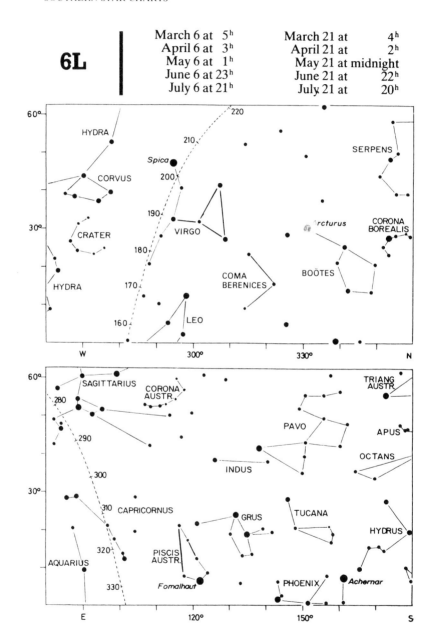

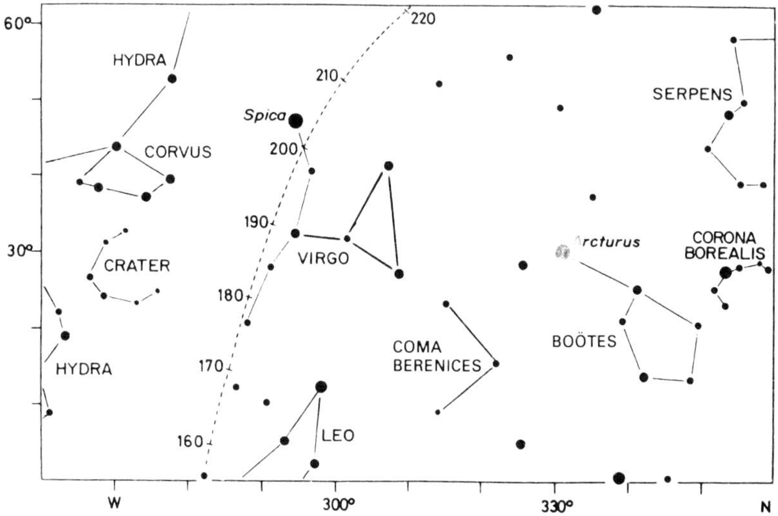

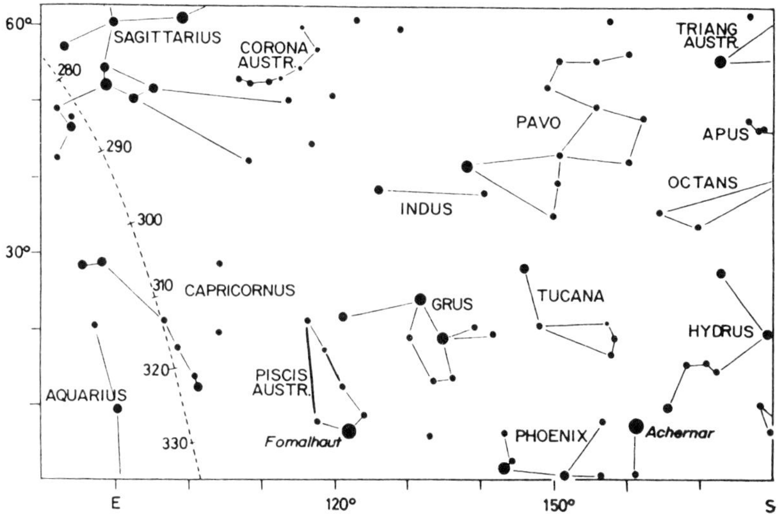

March 6 at 5^h	March 21 at 4^h	
April 6 at 3^h	April 21 at 2^h	
May 6 at 1^h	May 21 at midnight	**6R**
June 6 at 23^h	June 21 at 22^h	
July 6 at 21^h	July 21 at 20^h	

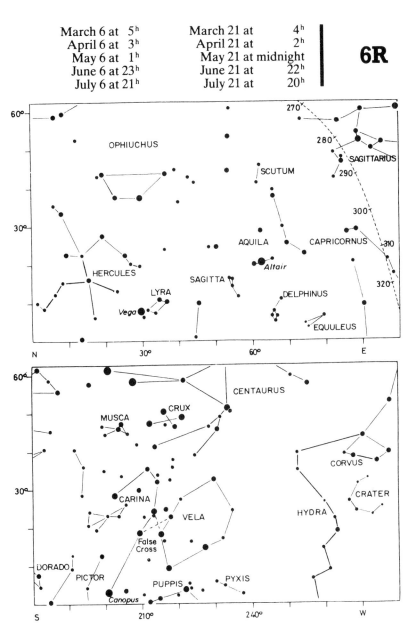

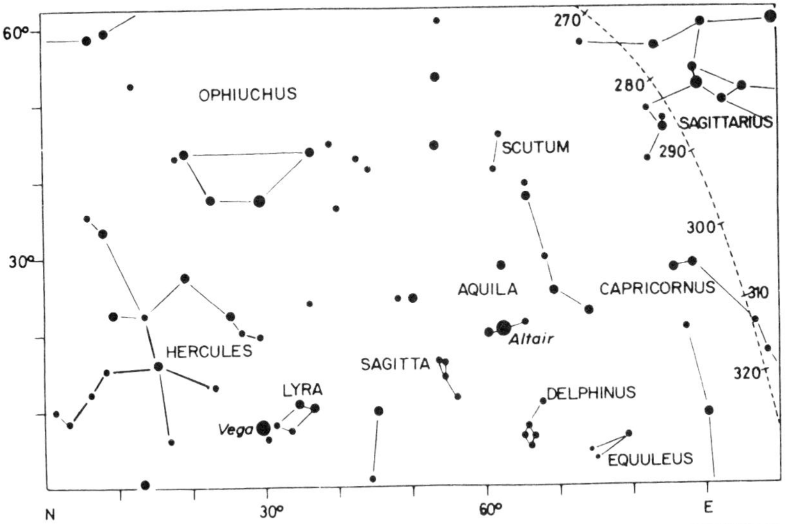

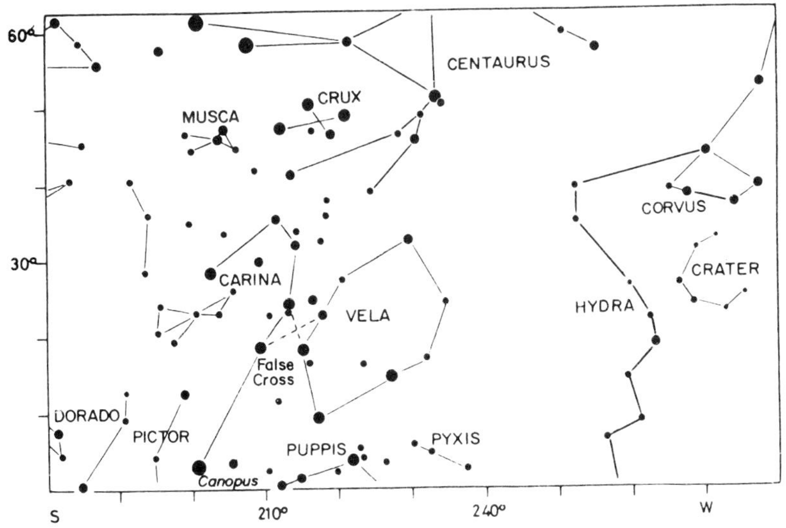

7L

April 6 at 5h	April 21 at 4h
May 6 at 3h	May 21 at 2h
June 6 at 1h	June 21 at midnight
July 6 at 23h	July 21 at 22h
August 6 at 21h	August 21 at 20h

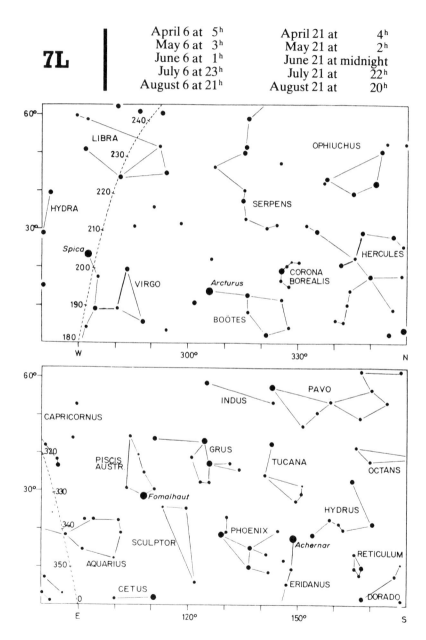

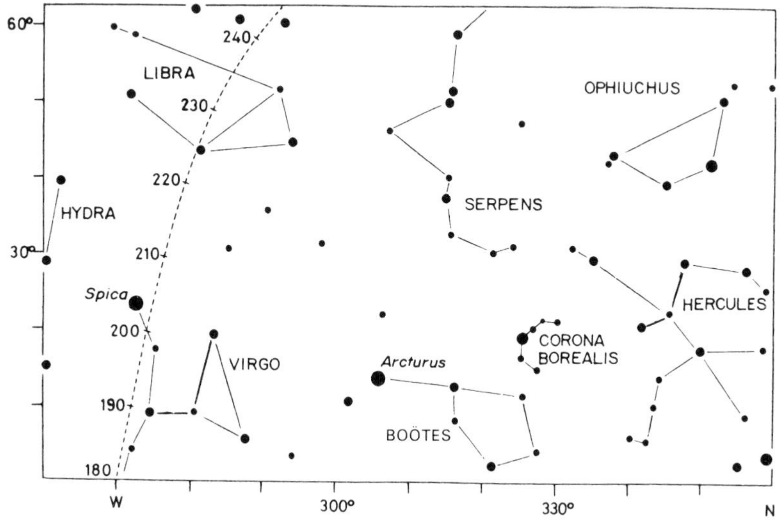

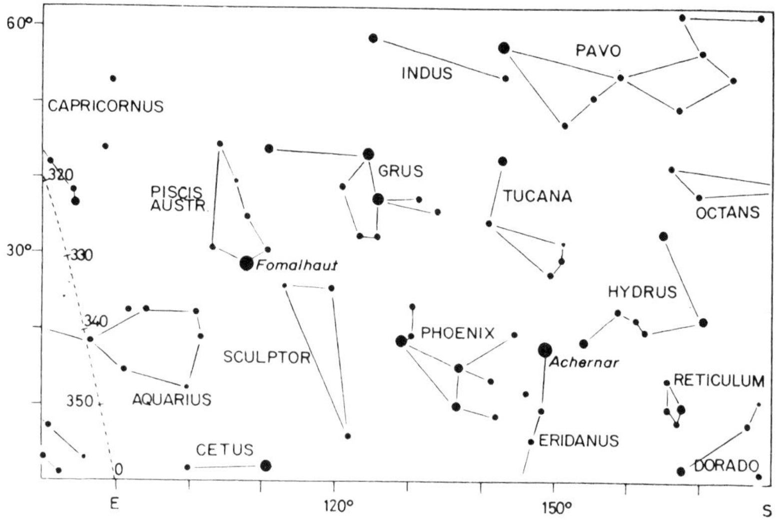

April 6 at 5ʰ	April 21 at 4ʰ	**7R**
May 6 at 3ʰ	May 21 at 2ʰ	
June 6 at 1ʰ	June 21 at midnight	
July 6 at 23ʰ	July 21 at 22ʰ	
August 6 at 21ʰ	August 21 at 20ʰ	

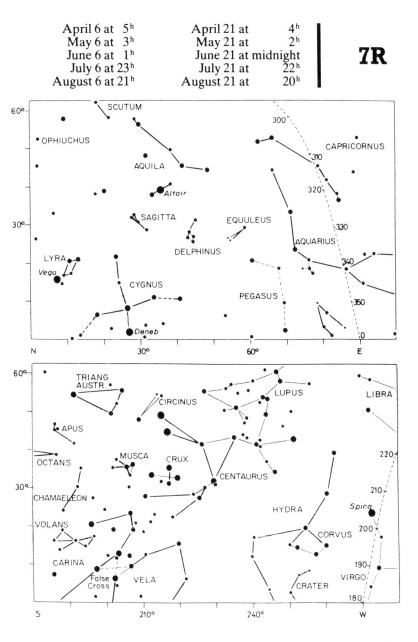

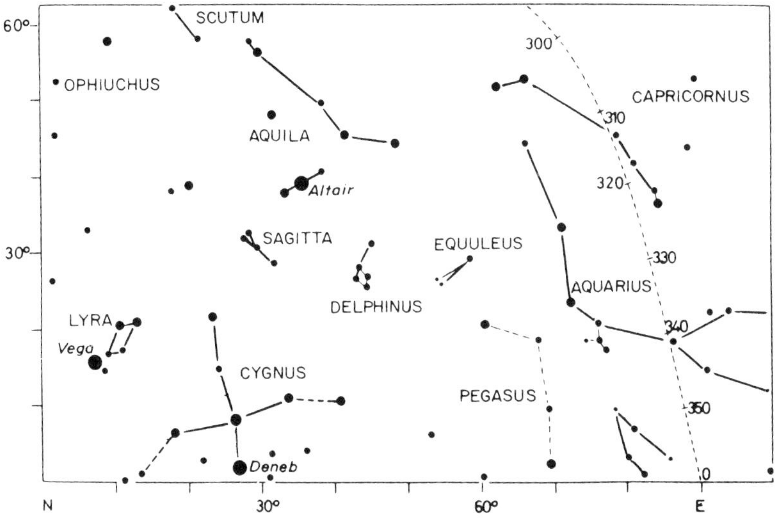

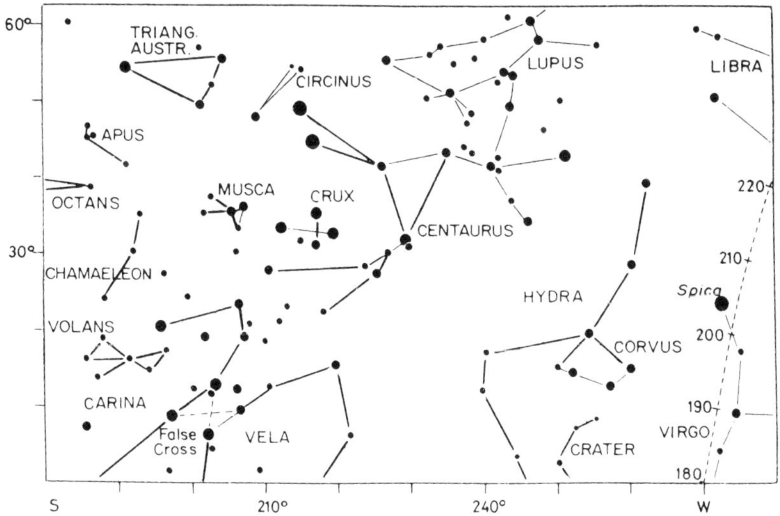

8L

May 6 at 5h	May 21 at 4h
June 6 at 3h	June 21 at 2h
July 6 at 1h	July 21 at midnight
August 6 at 23h	August 21 at 22h
September 6 at 21h	September 21 at 20h

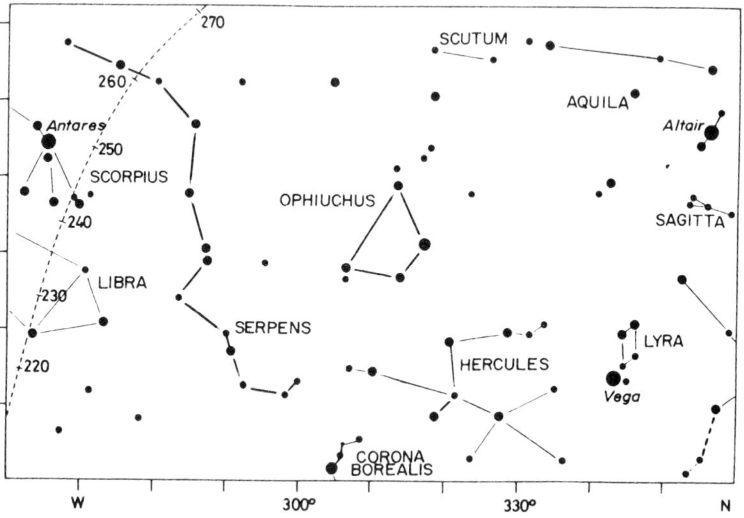

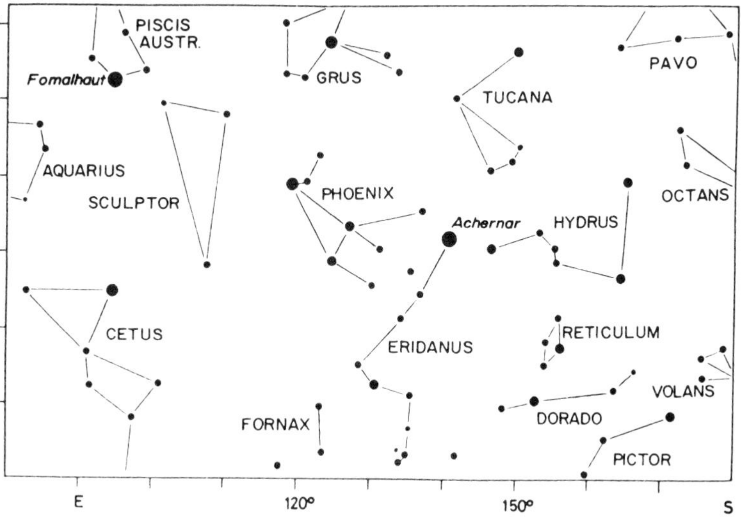

May 6 at 5h	May 21 at 4h	**8R**
June 6 at 3h	June 21 at 2h	
July 6 at 1h	July 21 at midnight	
August 6 at 23h	August 21 at 22h	
September 6 at 21h	September 21 at 20h	

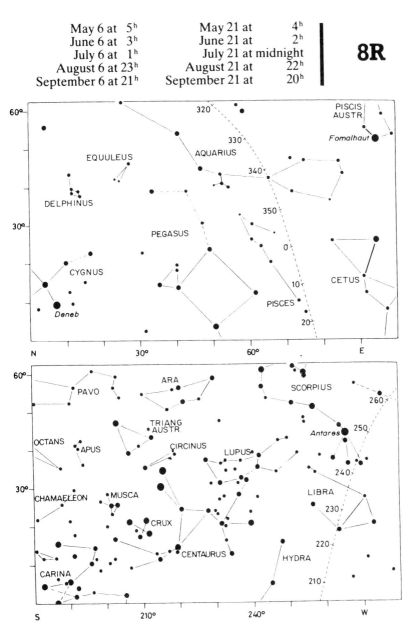

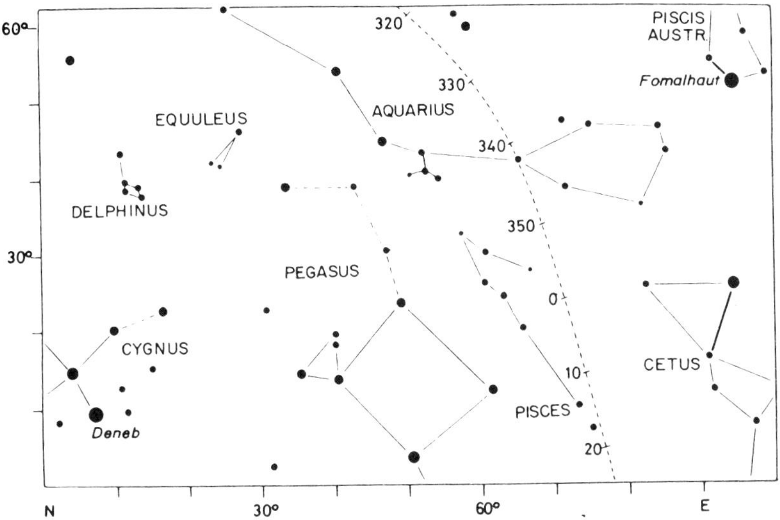

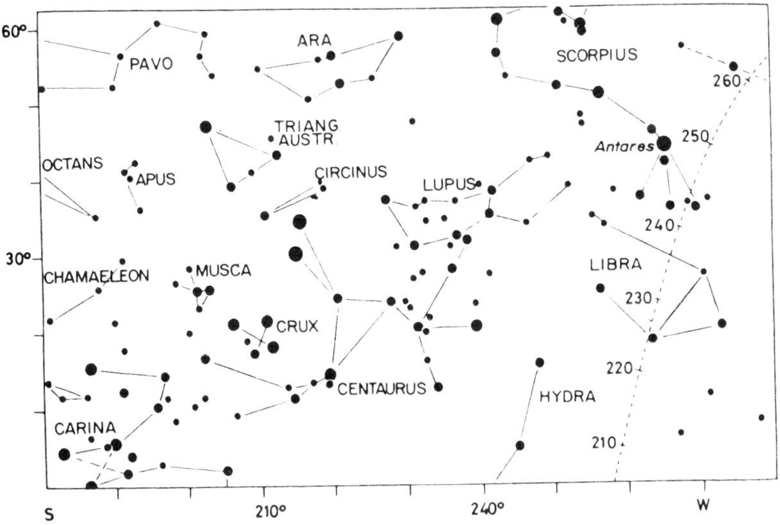

9L

June 6 at 5h	June 21 at 4h
July 6 at 3h	July 21 at 2h
August 6 at 1h	August 21 at midnight
September 6 at 23h	September 21 at 22h
October 6 at 21h	October 21 at 20h

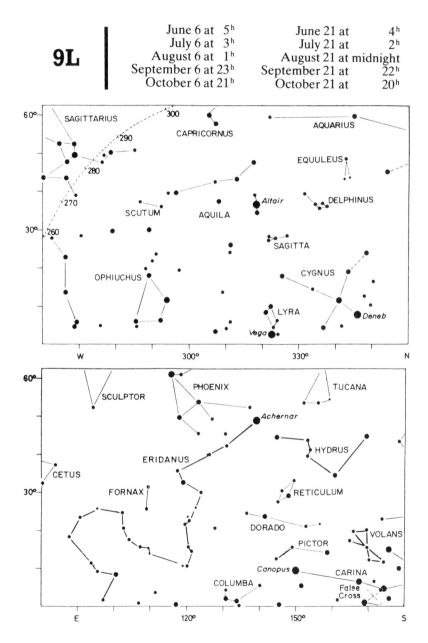

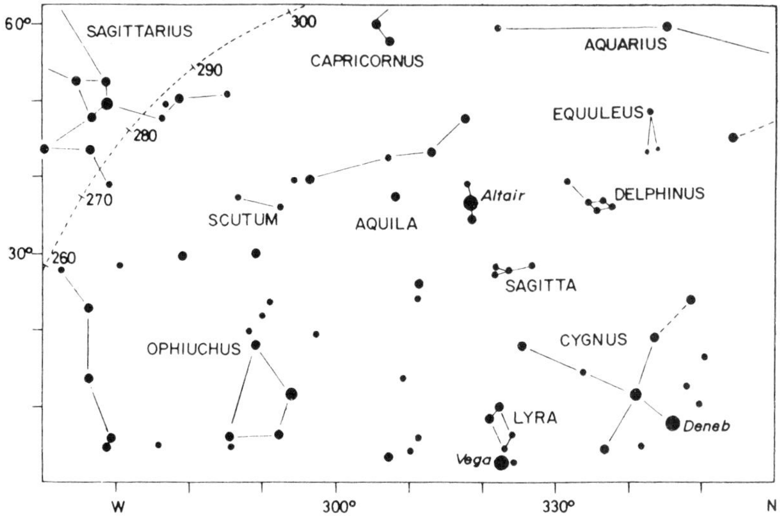

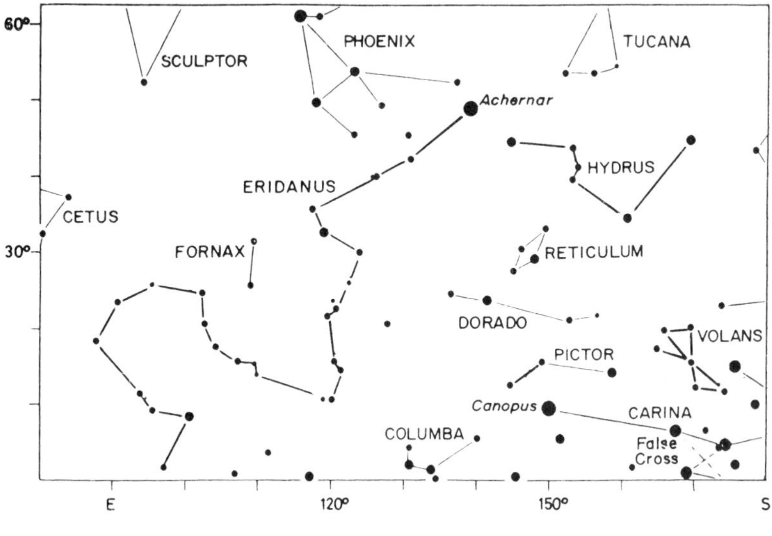

June 6 at	5^h	June 21 at	4^h
July 6 at	3^h	July 21 at	2^h
August 6 at	1^h	August 21 at	midnight
September 6 at	23^h	September 21 at	22^h
October 6 at	21^h	October 21 at	20^h

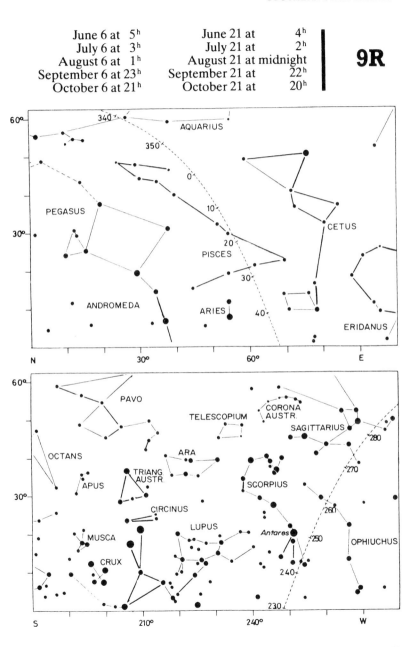

9R

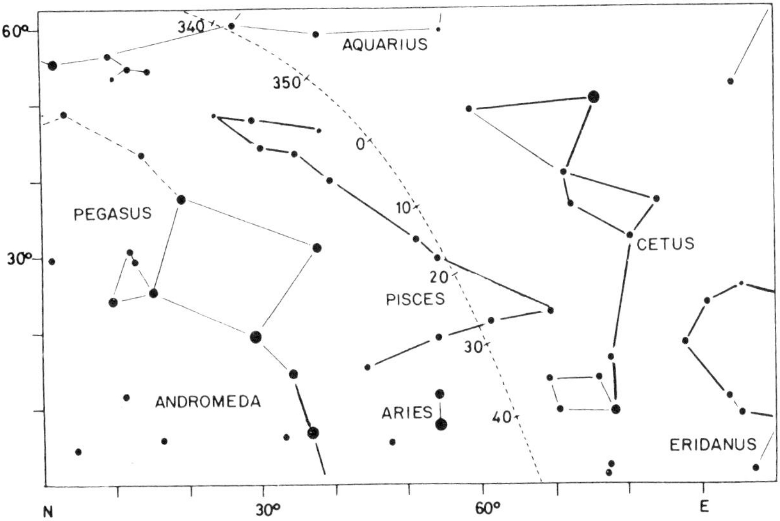

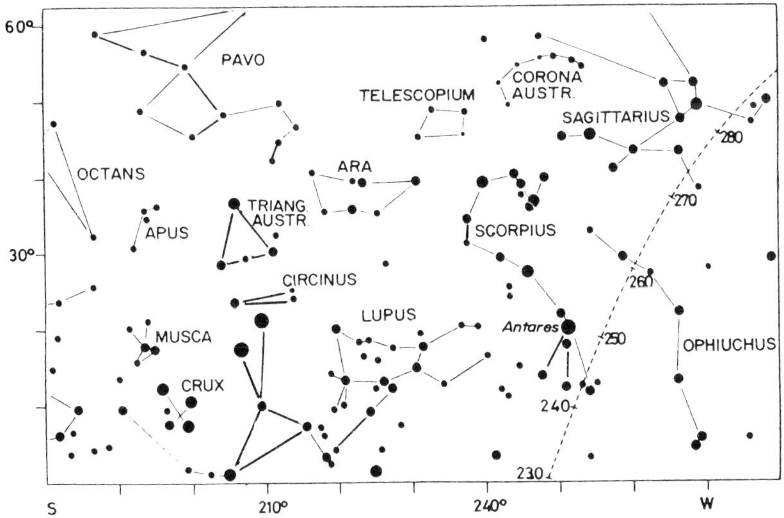

10L

July 6 at 5^h	July 21 at	4^h
August 6 at 3^h	August 21 at	2^h
September 6 at 1^h	September 21 at	midnight
October 6 at 23^h	October 21 at	22^h
November 6 at 21^h	November 21 at	20^h

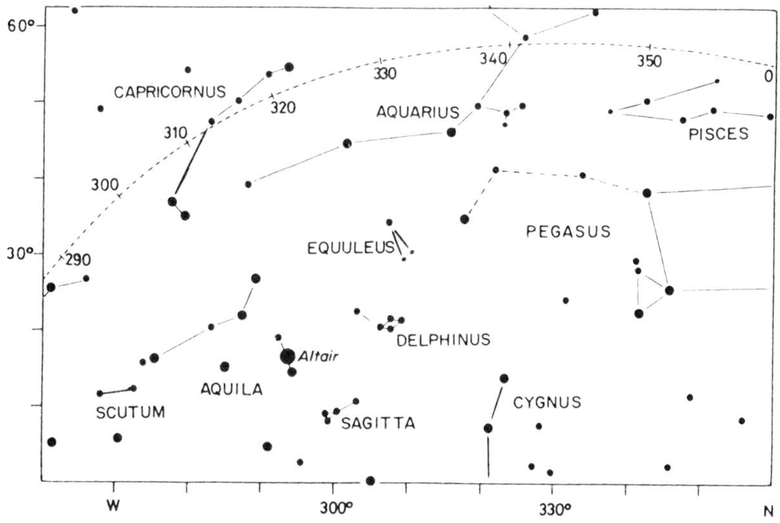

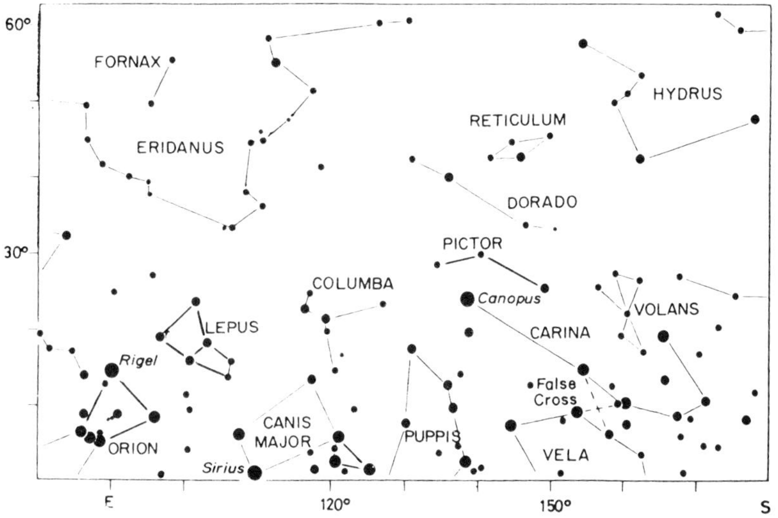

July 6 at 5^h	July 21 at 4^h	**10R**
August 6 at 3^h	August 21 at 2^h	
September 6 at 1^h	September 21 at midnight	
October 6 at 23^h	October 21 at 22^h	
November 6 at 21^h	November 21 at 20^h	

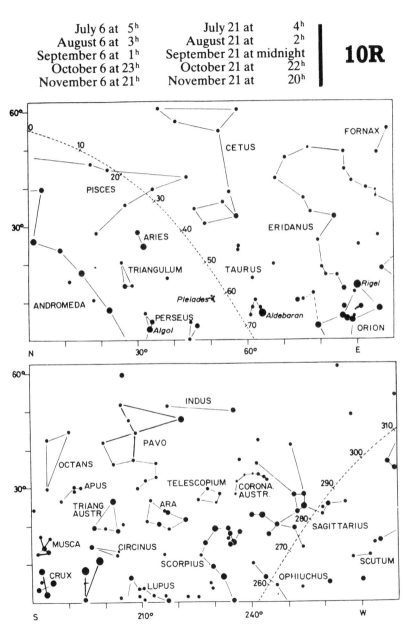

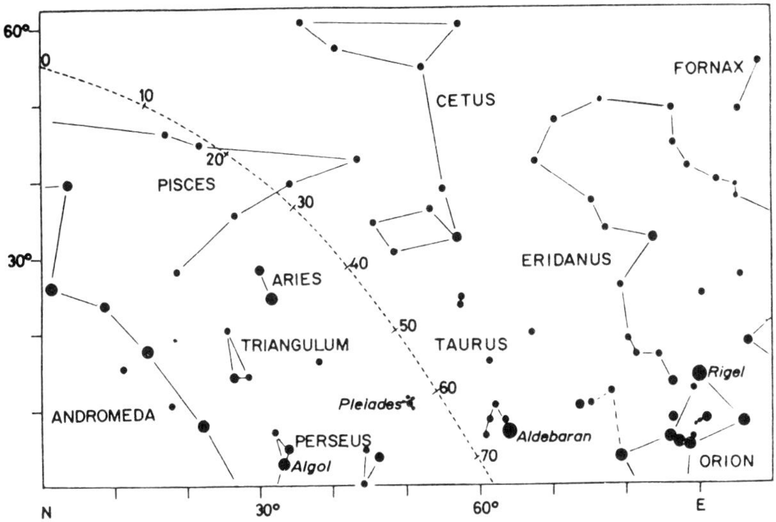

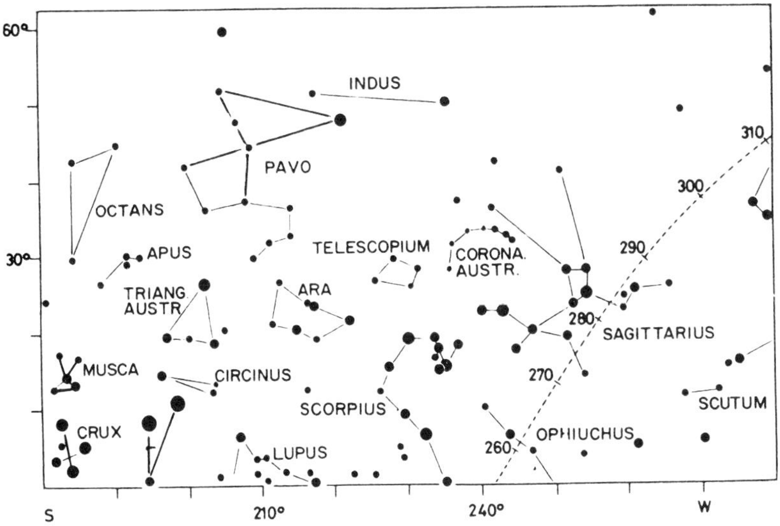

11L

August 6 at 5ʰ	August 21 at 4ʰ
September 6 at 3ʰ	September 21 at 2ʰ
October 6 at 1ʰ	October 21 at midnight
November 6 at 23ʰ	November 21 at 22ʰ
December 6 at 21ʰ	December 21 at 20ʰ

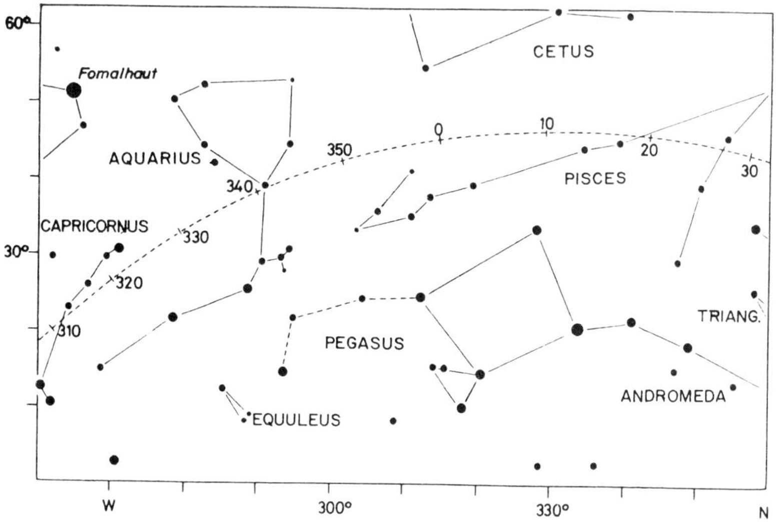

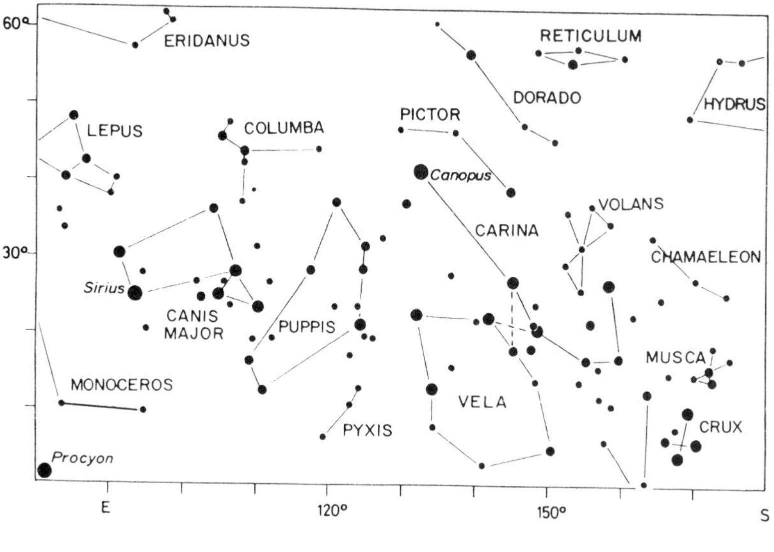

August 6 at 5h	August 21 at 4h	**11R**
September 6 at 3h	September 21 at 2h	
October 6 at 1h	October 21 at midnight	
November 6 at 23h	November 21 at 22h	
December 6 at 21h	December 21 at 20h	

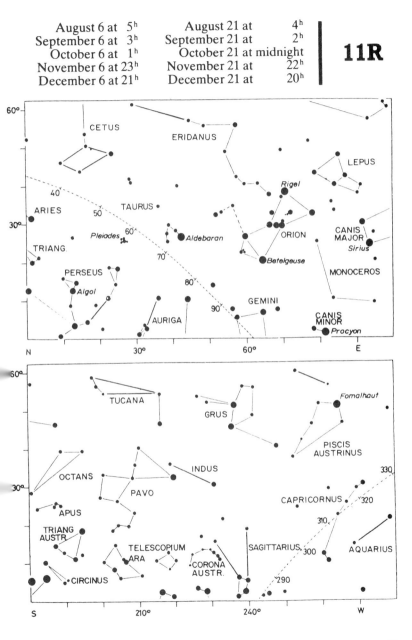

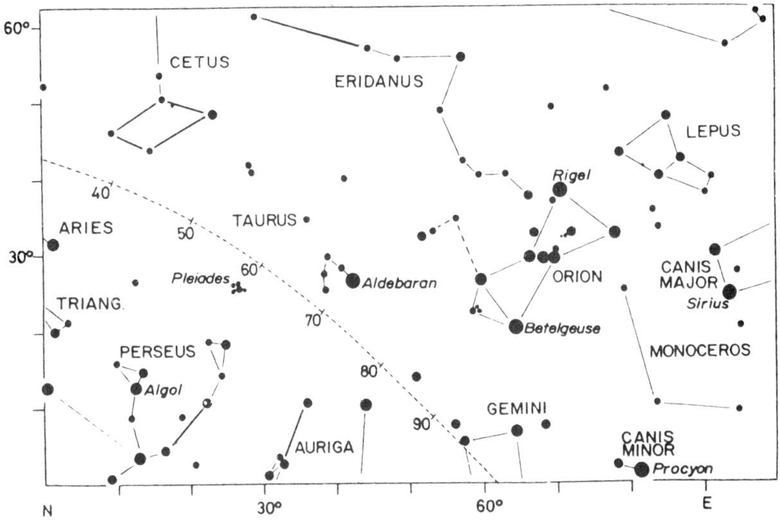

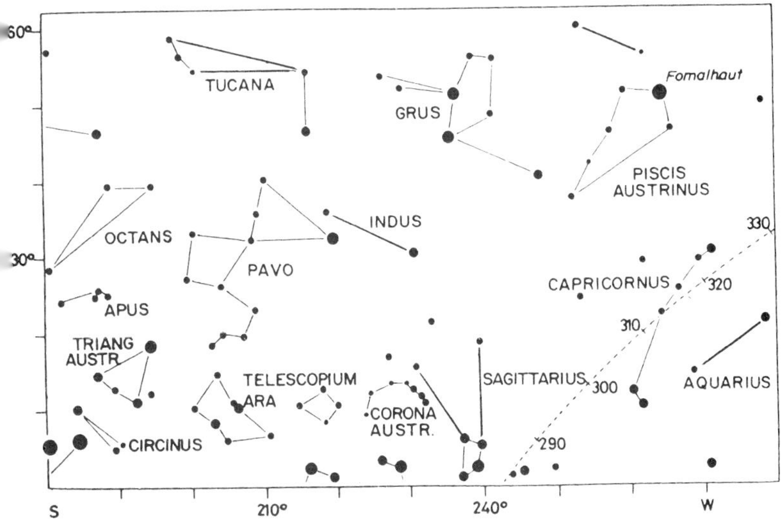

12L

September 6 at 5^h	September 21 at 4^h
October 6 at 3^h	October 21 at 2^h
November 6 at 1^h	November 21 at midnight
December 6 at 23^h	December 21 at 22^h
January 6 at 21^h	January 21 at 20^h

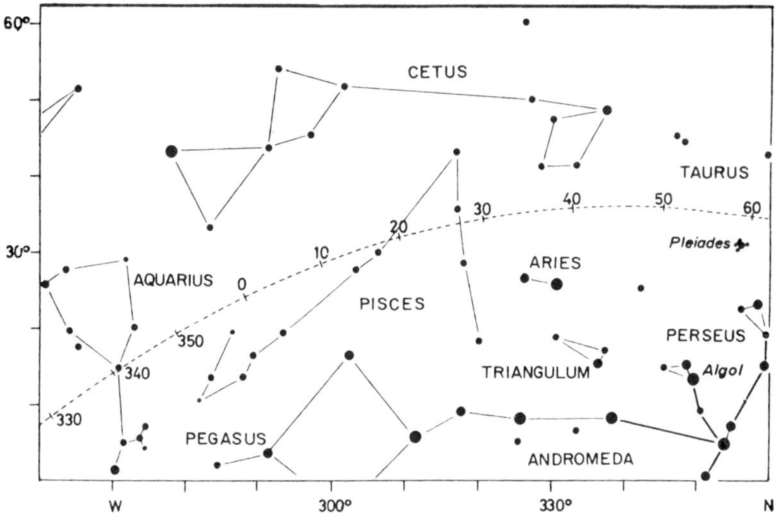

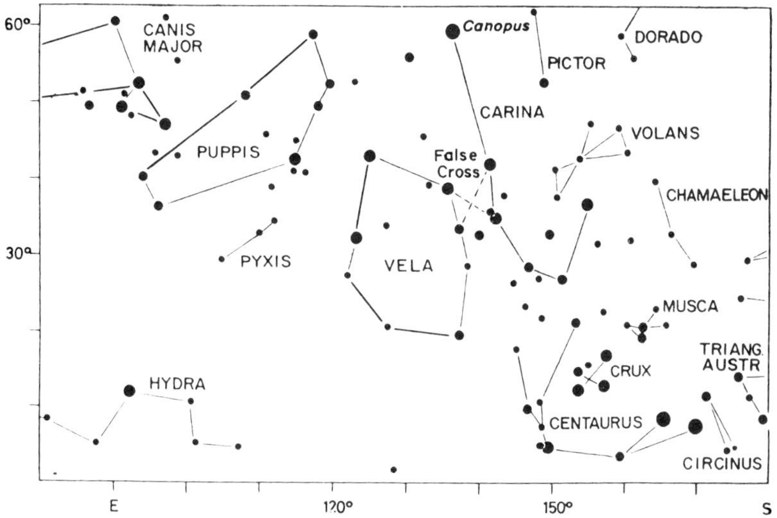

September 6 at 5h	September 21 at 4h		**12R**
October 6 at 3h	October 21 at 2h		
November 6 at 1h	November 21 at midnight		
December 6 at 23h	December 21 at 22h		
January 6 at 21h	January 21 at 20h		

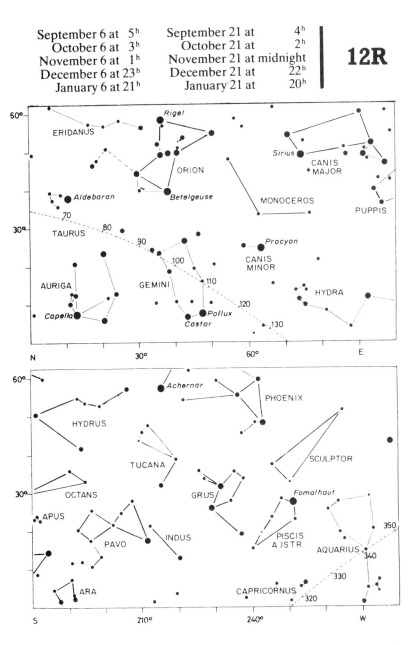

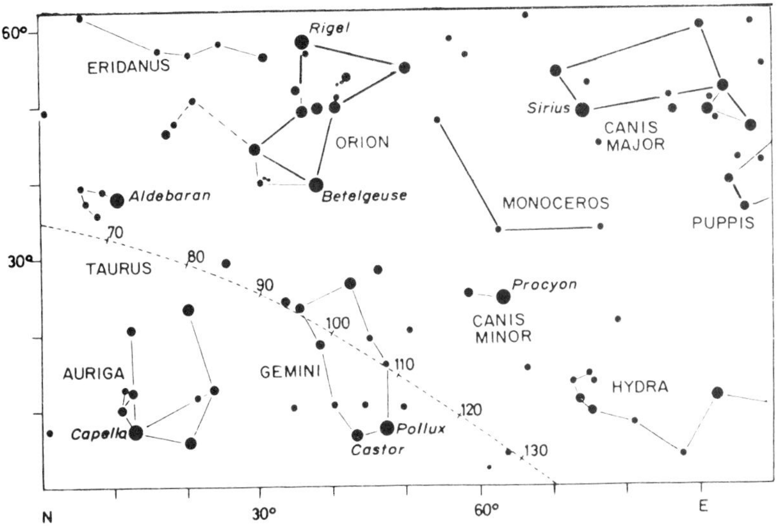

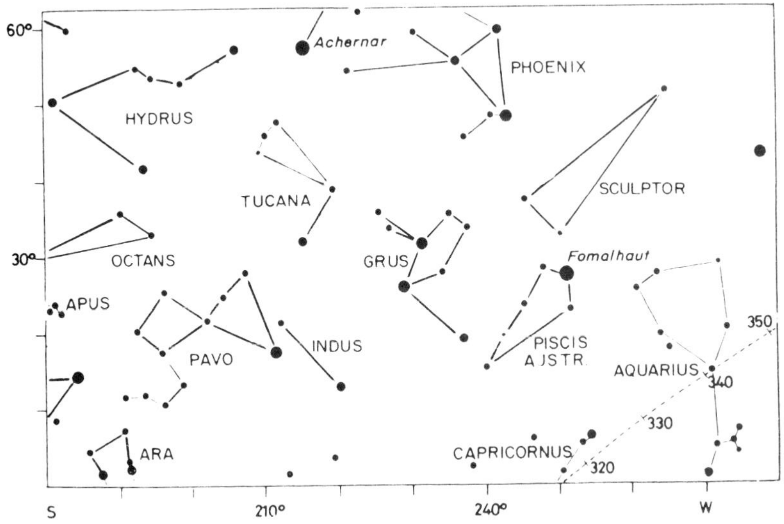

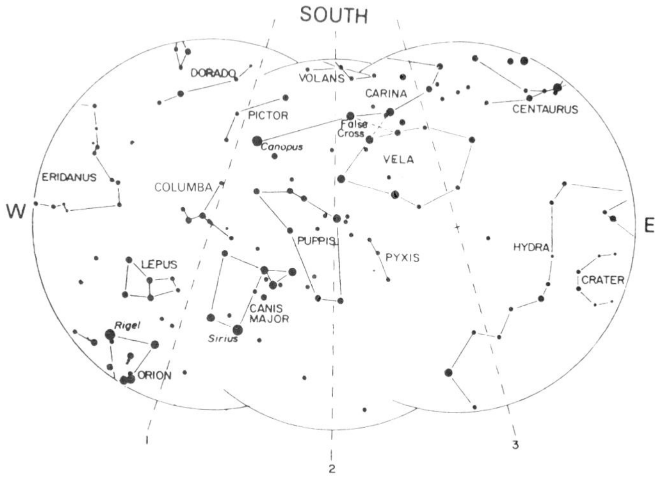

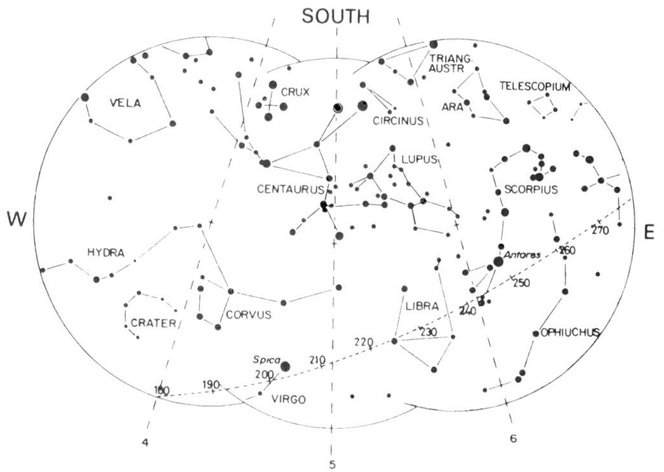

Southern Hemisphere Overhead Stars

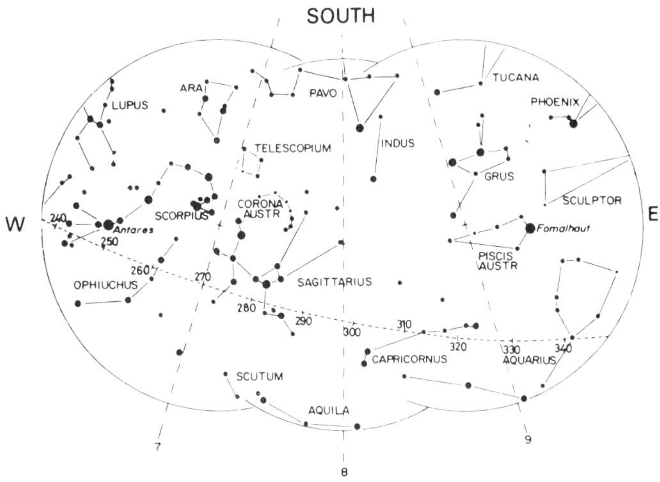

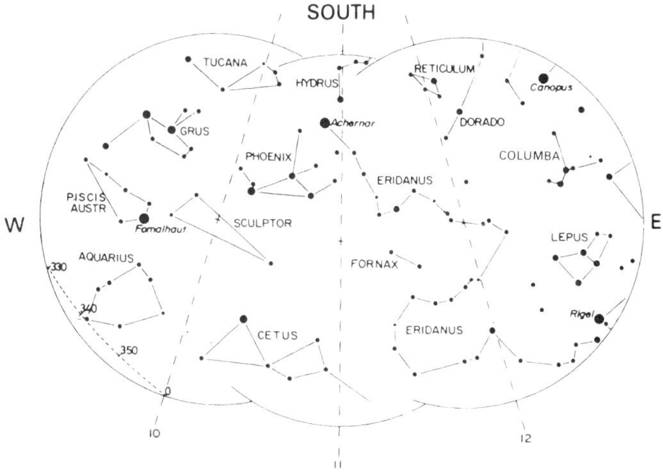

Southern Hemisphere Overhead Stars

The Planets and the Ecliptic

The paths of the planets about the Sun all lie close to the plane of the ecliptic, which is marked for us in the sky by the apparent path of the Sun among the stars, and is shown on the star charts by a broken line. The Moon and planets will always be found close to this line, never departing from it by more than about 7°. Thus the planets are most favourably placed for observation when the ecliptic is well displayed, and this means that it should be as high in the sky as possible. This avoids the difficulty of finding a clear horizon, and also overcomes the problem of atmospheric absorption, which greatly reduces the light of the stars. Thus a star at an altitude of 10° suffers a loss of 60 per cent of its light, which corresponds to a whole magnitude; at an altitude of only 4°, the loss may amount to two magnitudes.

The position of the ecliptic in the sky is therefore of great importance, and since it is tilted at about 23½° to the Equator, it is only at certain times of the day or year that it is displayed to the best advantage. It will be realized that the Sun (and therefore the ecliptic) is at its highest in the sky at noon in midsummer, and at its lowest at noon in midwinter. Allowing for the daily motion of the sky, it follows that the ecliptic is highest at midnight in winter, at sunset in the spring, at noon in summer and at sunrise in the autumn. Hence these are the best times to see the planets. Thus, if Venus is an evening object in the western sky after sunset, it will be seen to best advantage if this occurs in the spring, when the ecliptic is high in the sky and slopes down steeply to the horizon. This means that the planet is not only higher in the sky, but will remain for a much longer period above the horizon. For similar reasons, a morning object will be seen at its best on autumn mornings before sunrise, when the ecliptic is high in the east. The outer planets, which can come to opposition (i.e. opposite the Sun), are best seen when opposition occurs in the winter months, when the ecliptic is high in the sky at midnight.

The seasons are reversed in the Southern Hemisphere, spring beginning at the September Equinox, when the Sun crosses the Equator on its way south, summer beginning at the December Solstice, when the Sun is highest in the southern sky, and so on.

Thus, the times when the ecliptic is highest in the sky, and therefore best placed for observing the planets, may be summarized as follows:

	Midnight	*Sunrise*	*Noon*	*Sunset*
Northern lats.	December	September	June	March
Southern lats.	June	March	December	September

In addition to the daily rotation of the celestial sphere from east to west, the planets have a motion of their own among the stars. The apparent movement is generally *direct*, i.e. to the east, in the direction of increasing longitude, but for a certain period (which depends on the distance of the planet) this apparent motion is reversed. With the outer planets this *retrograde* motion occurs about the time of opposition. Owing to the different inclination of the orbits of these planets, the actual effect is to cause the apparent path to form a loop, or sometimes an S-shaped curve. The same effect is present in the motion of the inferior planets, Mercury and Venus, but it is not so obvious, since it always occurs at the time of inferior conjunction.

The *inferior planets*, Mercury and Venus, move in smaller orbits than that of the Earth, and so are always seen near the Sun. They are most obvious at the times of greatest angular distance from the Sun (greatest elongation), which may reach 28° for Mercury, and 47° for Venus. They are seen as evening objects in the western sky after sunset (at eastern elongations) or as morning objects in the eastern sky before sunrise (at western elongations). The succession of phenomena, conjunctions and elongations, always follows the same order, but the intervals between them are not equal. Thus, if either planet is moving round the far side of its orbit its motion will be to the east, in the same direction in which the Sun appears to be moving. It therefore takes much longer for the planet to overtake the Sun – that is, to come to superior conjunction – than it does when moving round to inferior conjunction, between Sun and Earth. The intervals given in the following table are average values; they remain fairly constant in the case of Venus, which travels in an almost circular orbit. In the case of Mercury, however, conditions vary widely because of the great eccentricity and inclination of the planet's orbit.

		Mercury	*Venus*
Inferior Conjunction	to Elongation West	22 days	72 days
Elongation West	to Superior Conjunction	36 days	220 days
Superior Conjunction	to Elongation East	36 days	220 days
Elongation East	to Inferior Conjunction	22 days	72 days

The greatest brilliancy of Venus always occurs about 36 days before or after inferior conjunction. This will be about a month *after* greatest eastern elongation (as an evening object), or a month *before* greatest western elongation (as a morning object). No such rule can be given for Mercury, because its distance from the Earth and the Sun can vary over a wide range.

Mercury is not likely to be seen unless a clear horizon is available. It is seldom as much as 10° above the horizon in the twilight sky in northern latitudes, but this figure is often exceeded in the Southern Hemisphere. This favourable condition arises because the maximum elongation of 28° can occur only when the planet is at aphelion (farthest from the Sun), and it then lies well south of the Equator. Northern observers must be content with smaller elongations, which may be as little as 18° at perihelion. In general, it may be said that the most favourable times for seeing Mercury as an evening object will be in spring, some days before greatest eastern elongation; in autumn, it may be seen as a morning object some days after greatest western elongation.

Venus is the brightest of the planets and may be seen on occasions in broad daylight. Like Mercury, it is alternately a morning and an evening object, and it will be highest in the sky when it is a morning object in autumn, or an evening object in spring. The phenomena of Venus given in the table on p. 67 can occur only in the months of January, April, June, August and November, and it will be realized that they do not all lead to favourable apparitions of the planet. In fact, Venus is to be seen at its best as an evening object in northern latitudes when eastern elongation occurs in June. The planet is then well north of the Sun in the preceding spring months, and is a brilliant object in the evening sky over a long period. In the Southern Hemisphere a November elongation is best. For similar reasons, Venus gives a prolonged display as a morning object in the months following western elongation in November (in northern latitudes) or in June (in the Southern Hemisphere).

The *superior planets*, which travel in orbits larger than that of the Earth, differ from Mercury and Venus in that they can be seen opposite the Sun in the sky. The superior planets are morning objects after conjunction with the Sun, rising earlier each day until they come to opposition. They will then be nearest to the Earth (and therefore at their brightest), and will be on the meridian at midnight, due south in northern latitudes, but due north in the Southern Hemisphere. After opposition they are evening objects, setting earlier each evening until they set in the west with the Sun at the next conjunction. The difference in brightness from one opposition to another is most noticeable in the case of Mars, whose distance from Earth can vary considerably and rapidly. The other superior planets are at such great distances that there is very little change in brightness from one opposition to the next. The effect of altitude is, however, of some importance, for at a December opposition in northern latitudes the planets will be among the stars of Taurus or Gemini, and can then be at an altitude of more than 60° in southern England. At a summer opposition, when the planet is in Sagittarius, it may only rise to about 15° above the southern horizon, and so makes a less impressive appearance. In the Southern Hemisphere the reverse conditions apply, a June opposition being the best, with the planet in Sagittarius at an altitude which can reach 80° above the northern horizon for observers in South Africa.

Mars, whose orbit is appreciably eccentric, comes nearest to the Earth at oppositions at the end of August. It may then be brighter even than Jupiter, but rather low in the sky in Aquarius for northern observers, though very well placed for those in southern latitudes. These favourable oppositions occur every fifteen or seventeen years (1988, 2003, 2018), but in the Northern Hemisphere the planet is probably better seen at oppositions in the autumn or winter months, when it is higher in the sky. Oppositions of Mars occur at an average interval of 780 days, and during this time the planet makes a complete circuit of the sky.

Jupiter is always a bright planet, and comes to opposition a month later each year, having moved, roughly speaking, from one Zodiacal constellation to the next.

Saturn moves much more slowly than Jupiter, and may remain in the same constellation for several years. The brightness of Saturn depends on the aspects of its rings, as well as on the distance from Earth and Sun. The Earth passed through the plane of Saturn's rings in 1995 and 1996, when they appeared edge-on; we shall next see them at maximum opening, and Saturn at its brightest, around 2002. The rings will next appear edge-on in 2009.

Uranus, *Neptune* and *Pluto* are hardly likely to attract the attention of observers without adequate instruments.

Phases of the Moon, 1998

New Moon		*First Quarter*		*Full Moon*		*Last Quarter*	
	d h m		*d h m*		*d h m*		*d h m*
		Jan.	5 14 18	Jan.	12 17 24	Jan.	20 19 40
Jan.	28 06 01	Feb.	3 22 53	Feb.	11 10 23	Feb.	19 15 27
Feb.	26 17 26	Mar.	5 08 41	Mar.	13 04 34	Mar.	21 07 38
Mar.	28 03 14	Apr.	3 20 18	Apr.	11 22 23	Apr.	19 19 53
Apr.	26 11 41	May	3 10 04	May	11 14 29	May	19 04 35
May	25 19 32	June	2 01 45	June	10 04 18	June	17 10 38
June	24 03 50	July	1 18 43	July	9 16 01	July	16 15 13
July	23 13 44	July	31 12 05	Aug.	8 02 10	Aug.	14 19 48
Aug.	22 02 03	Aug.	30 05 06	Sept.	6 11 21	Sept.	3 01 58
Sept.	20 17 02	Sept.	28 21 11	Oct.	5 20 12	Oct.	12 11 11
Oct.	20 10 09	Oct.	28 11 46	Nov.	4 05 18	Nov.	11 00 28
Nov.	19 04 27	Nov.	27 00 23	Dec.	3 15 19	Dec.	10 17 54
Dec.	18 22 42	Dec.	26 10 46				

All times are GMT.

Longitudes of the Sun, Moon and Planets in 1998

		Sun	Moon	Venus	Mars	Jupiter	Saturn
		°	°	°	°	°	°
January	6	285	21	302	315	323	14
	21	301	213	294	326	327	15
February	6	317	73	288	339	330	16
	21	332	258	292	351	334	17
March	6	345	83	301	1	337	19
	21	0	266	313	12	341	20
April	6	16	131	330	25	344	22
	21	31	315	345	36	347	24
May	6	45	164	2	47	350	26
	21	60	353	19	57	353	28
June	6	75	208	38	69	355	30
	21	89	47	55	79	357	31
July	6	104	240	74	90	358	32
	21	118	86	91	99	358	33
August	6	133	287	111	110	357	34
	21	148	135	129	120	356	34
September	6	163	337	149	130	354	33
	21	178	181	167	139	352	33
October	6	193	15	186	149	351	32
	21	207	214	204	158	349	30
November	6	223	69	225	168	348	29
	21	238	258	243	176	348	28
December	6	254	106	263	185	349	27
	21	269	292	281	193	350	27

Longitude of *Uranus* 310°
Neptune 301°

Moon: Longitude of ascending node
Jan. 1: 164° Dec. 31: 144°

Mercury moves so quickly among the stars that it is not possible to indicate its position on the star charts at convenient intervals. The monthly notes must be consulted for the best times at which the planet may be seen.

The positions of the other planets are given in the table on p. 71. This gives the apparent longitudes on dates which correspond to those of the star charts, and the position of the planet may at once be found near the ecliptic at the given longitude.

Examples

In the Southern Hemisphere two planets are seen in the eastern morning sky in late April. Identify them.

> The Southern Star Chart 7L shows the eastern sky at April 21^{d} 04^{h} and shows longitudes 320° to 0°. Reference to the table on p. 71 gives the longitude of Venus as 345° and that of Jupiter as 347°. Thus these planets are to be found in the eastern sky, and the brighter one is Venus.

The positions of the Sun and Moon can be plotted on the star maps in the same manner as for the planets. The average daily motion of the Sun is 1°, and of the Moon 13°. For the Moon an indication of its position relative to the ecliptic may be obtained from a consideration of its longitude relative to that of the ascending node. The latter changes only slowly during the year, as will be seen from the values given on p. 71. Let us denote by d the difference in longitude between the Moon and its ascending node. Then if $d = 0°$, 180° or 360° the Moon is on the ecliptic. If $d = 90°$ the Moon is 5° north of the ecliptic, and if $d = 270°$ the Moon is 5° south of the ecliptic.

On February 6 the Moon's longitude is given as 73°, and the longitude of the node is found by interpolation to be about 162°. Thus $d = 271°$, and the Moon is about 5° south of the ecliptic. Its position may be plotted on Northern Star Charts 1R, 2R, 3L, 3R, 4L, 9R, 10L, 10R, 11L and 12L, and on Southern Star Charts 1L, 2L, 11R and 12R.

Events in 1998

ECLIPSES

There will be two eclipses, both of the Sun.

February 26: total eclipse of the Sun – the Americas.
August 21–22: annular eclipse of the Sun – SE Asia, Australasia.

THE PLANETS

Mercury may be seen more easily from northern latitudes in the evenings about the time of greatest eastern elongation (March 20), and in the mornings around greatest western elongation (August 31). In the Southern Hemisphere the corresponding most favourable dates are around January 6 (mornings) and November 11 (evenings).

Venus is visible in the evenings for the first ten days of January. It is visible in the mornings for the last ten days of January and until late in September. It is again visible in the evenings for the second half of December.

Mars is at conjunction on May 12.

Jupiter is at opposition on September 16.

Saturn is at opposition on October 23.

Uranus is at opposition on August 3.

Neptune is at opposition on July 23.

Pluto is at opposition on May 28.

JANUARY

New Moon: January 28 *Full Moon*: January 12

EARTH is at perihelion (nearest to the Sun) on January 4, at a distance of 147 million kilometres.

MERCURY attains its greatest western elongation (23°) on January 6. It is about as far south of the equator as the Sun is, and thus poorly placed for observation by those in northern temperate latitudes where it may be seen only for the first ten days of the month, very low above the south-eastern horizon, about the time of beginning of morning civil twilight. However, Mercury is much better situated for observation by those in the Southern Hemisphere, where it will best be seen above the east-south-eastern horizon about half an hour before sunrise throughout the month. During January, Mercury increases very slowly in brightness, its magnitude changing from −0.1 to −0.5.

VENUS, magnitude −4.3, is a brilliant object in the early evenings, low in the south-western sky after sunset for the first fortnight in January. It moves rapidly through inferior conjunction on January 16, and for the last two weeks of the month is visible shortly before sunrise, low above the east-south-eastern horizon. A keen observer might notice that Venus is not actually on the ecliptic. In fact at the end of January Venus is as much as 7°.2N of it. This is due to its proximity to the Earth; its heliocentric latitude is only +2°.9.

MARS is unsuitably placed for observation by those in northern temperate latitudes. Even for observers further south, Mars will be a difficult object to locate, low above the south-western horizon for a short while after the end of evening civil twilight. Mars will certainly be lost to view before the end of the month. Its path amongst the stars is shown in Figure 1.

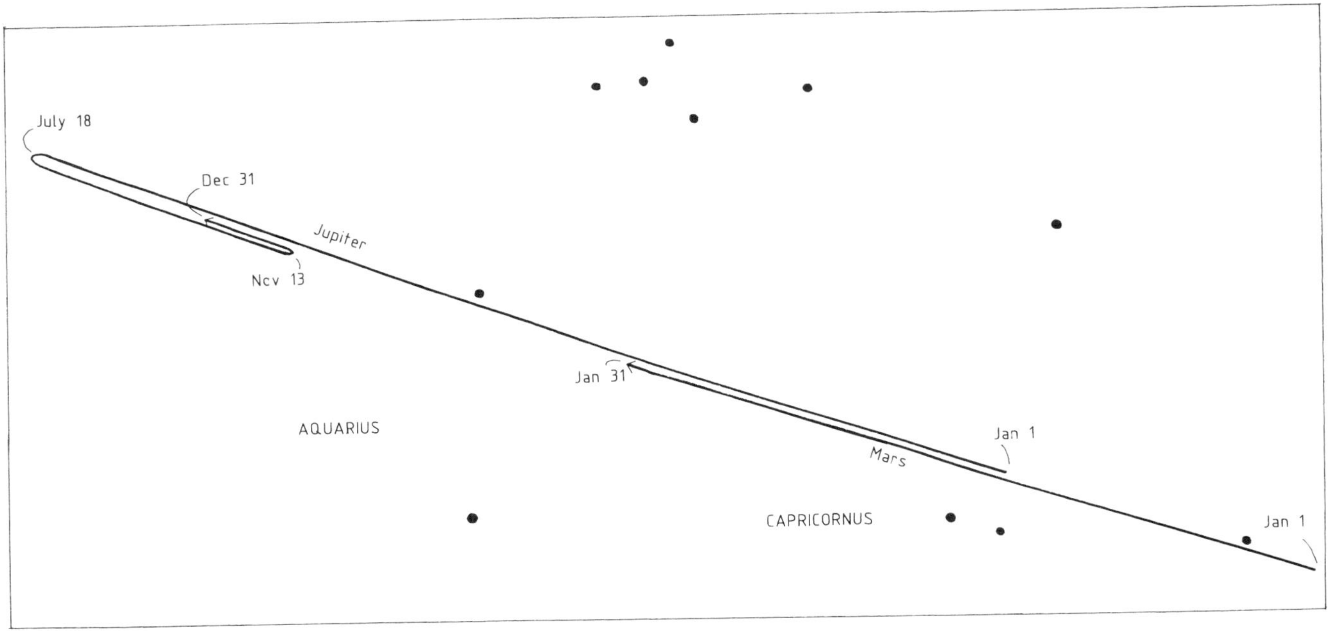

Figure 1. The path of Jupiter in 1998, and the path of Mars for January 1998.

JUPITER is an evening object, magnitude −2.0, visible in the south-western sky for a while after sunset. For observers in the the latitudes of the British Isles Jupiter is becoming a difficult object to locate by the end of the month. The path of Jupiter amongst the stars during 1998 is shown in Figure 1.

SATURN, magnitude +0.7, is an evening object in the western sky. Its movement amongst the stars during the year is shown in Figure 2, given with the notes for February.

FEBRUARY

New Moon: February 26 *Full Moon*: February 11

MERCURY, for observers in equatorial and southern latitudes only, may be detected as a difficult morning object low above the east-south-east horizon at the beginning of morning civil twilight, for the first week of the month. Its magnitude is −0.5. Mercury passes through superior conjunction on February 22 and therefore remains too close to the Sun for observation for the rest of the month.

VENUS is a brilliant object in the early mornings, completely dominating the eastern skies before dawn. Venus attains its greatest brilliancy (magnitude −4.6) on February 20. It is a beautiful sight in a small telescope at this time, as it exhibits a slender crescent phase. During the month the phase increases noticeably as the apparent diameter decreases.

MARS is too close to the Sun for observation.

JUPITER, magnitude −2.0, is becoming increasingly difficult to observe low in the south-western sky around the time of end of evening civil twilight. It is unlikely to be seen after the first week of the month since it passes through conjunction on February 23.

SATURN continues to be visible as an evening object in the western sky, magnitude +0.7. Its path amongst the stars is shown in Figure 2.

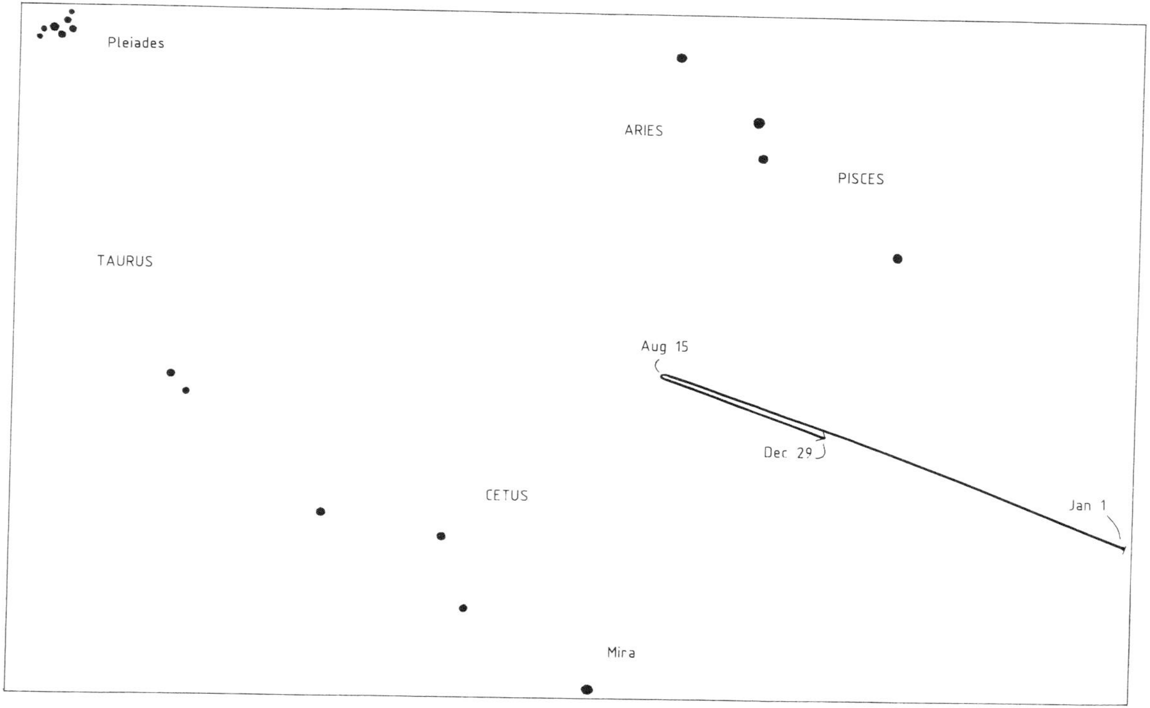

Figure 2. The path of Saturn in 1998.

MARCH

New Moon: March 28 *Full Moon*: March 13

Equinox: March 20

Summer Time in Great Britain and Northern Ireland commences on March 29.

MERCURY becomes an evening object after the first week of the month. As it is well north of the Sun in declination, observers in northern temperate latitudes will get the best view. Indeed, this evening apparition is the only suitable one of the year for northern observers.

For observers in northern temperate latitudes this will be the most favourable evening apparition of the year. Figure 3 shows, for

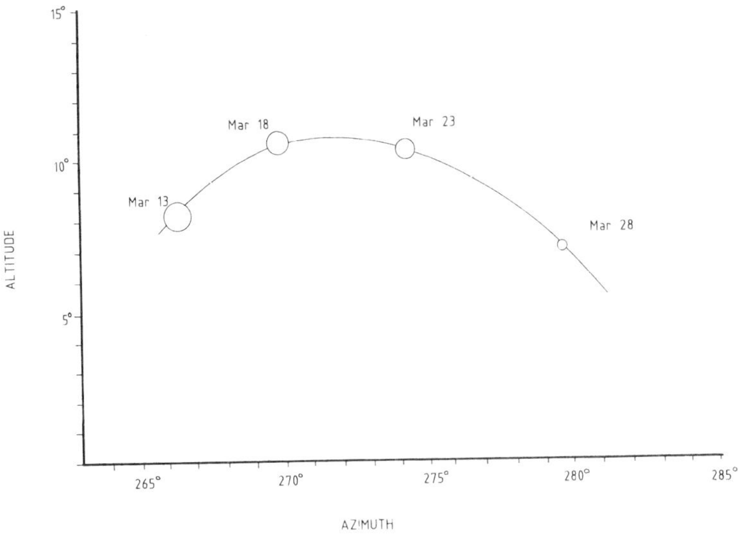

Figure 3. Evening apparition of Mercury, from latitude 52°N.

observers at latitude 52°N, the changes in azimuth (the true bearing from the north through east, south and west) and altitude of Mercury on successive evenings when the Sun is 6° below the horizon. This condition is known as the end of evening civil twilight, and for this latitude and time of year occurs about 35 minutes after sunset. The changes in the brightness of the planet are indicated by the relative sizes of the circles marking Mercury's position at five-day intervals. Mercury is at its brightest before it reaches greatest eastern elongation (19°) on March 20. Mercury's magnitude fades rapidly during the last week of March so that the planet will be increasingly difficult to detect, low above the western horizon, in the twilight sky.

VENUS is still a brilliant object in the morning skies, at magnitude −4.5. It is south of the celestial equator, and observers in the British Isles will be able to see it only for about an hour before sunrise, low above the east-south-eastern horizon. Observers in the southern hemisphere are more fortunate, enjoying a visibility period of around three hours before dawn. Venus is at greatest western elongation (47°) on March 27.

MARS is unsuitably placed for observation.

JUPITER, magnitude −2.0, emerges from the morning twilight around the middle of the month, for observers in equatorial and southern latitudes, when it may be seen low above the south-eastern horizon before twilight inhibits observation. Observers at the latitudes of the British Isles will not be able to see the planet until April.

SATURN, magnitude +0.7, is still an evening object in the western sky. For observers in the British Isles it is becoming increasingly difficult to detect as it moves closer to the Sun, and they are unlikely to see it after the middle of the month.

APRIL

New Moon: April 26 *Full Moon*: April 11

MERCURY passes through conjunction on April 6, and for the last few days of the month becomes visible to observers in equatorial regions and the Southern Hemisphere as a morning object in the east-north-eastern sky. The apparition lasts until the end of May; the increase in brightness of the planet during this period is quite marked, amounting to almost three magnitudes. Readers should refer to Figure 4, given with the notes for May.

VENUS, magnitude −4.2, continues to be visible as a brilliant object in the mornings before dawn. However, it is visible to observers in northern temperate latitudes only for a short while before sunset, low above the east-south-eastern horizon. Southern Hemisphere observers will continue to enjoy a three-hour period of visibility.

MARS is too close to the Sun for observation.

JUPITER, already observable to those in equatorial and southern latitudes, becomes visible to observers in northern temperate latitudes towards the end of the month. It will be seen above the east-south-eastern horizon for a short while before dawn, at magnitude −2.1.

SATURN is unsuitably placed for observation since it passes through conjunction on April 13.

MAY

New Moon: May 25 *Full Moon*: May 11

MERCURY, although not suitably placed for observation by those in northern temperate latitudes, continues to be visible as a morning object to those further south. For observers in southern latitudes this will be the most favourable morning apparition of the year. Figure 4 shows, for observers at latitude 35°S, the changes in azimuth (the true bearing from the north through east, south and west) and altitude of Mercury on successive evenings when the Sun is 6° below the horizon. This condition is known as the beginning of morning civil twilight, and for this latitude and time of year occurs about 30 minutes before sunrise. The changes in the brightness of the planet are indicated by the relative sizes of the circles marking Mercury's position at five-day intervals. Mercury is at its brightest after it reaches greatest western elongation (27°) on May 4. Note the proximity of Saturn around the middle of the month, Saturn being slightly fainter than Mercury.

VENUS is still a splendid object in the early morning before sunrise, at magnitude −4.1. Observers at northern temperate latitudes will find, however, that it is visible only for a short while before dawn, low above the eastern horizon.

MARS passes through conjunction on May 12 and is thus unsuitably placed for observation.

JUPITER, magnitude −2.2, is a morning object. For observers at the latitudes of the British Isles it is visible only for a short while, low in the south-eastern sky before the lengthening morning twilight inhibits observation. Further south, observers will see Jupiter as a conspicuous object in the south-eastern sky for several hours before dawn.

SATURN, for observers at the latitudes of the British Isles, remains unsuitably placed for observation due to the increasing length of

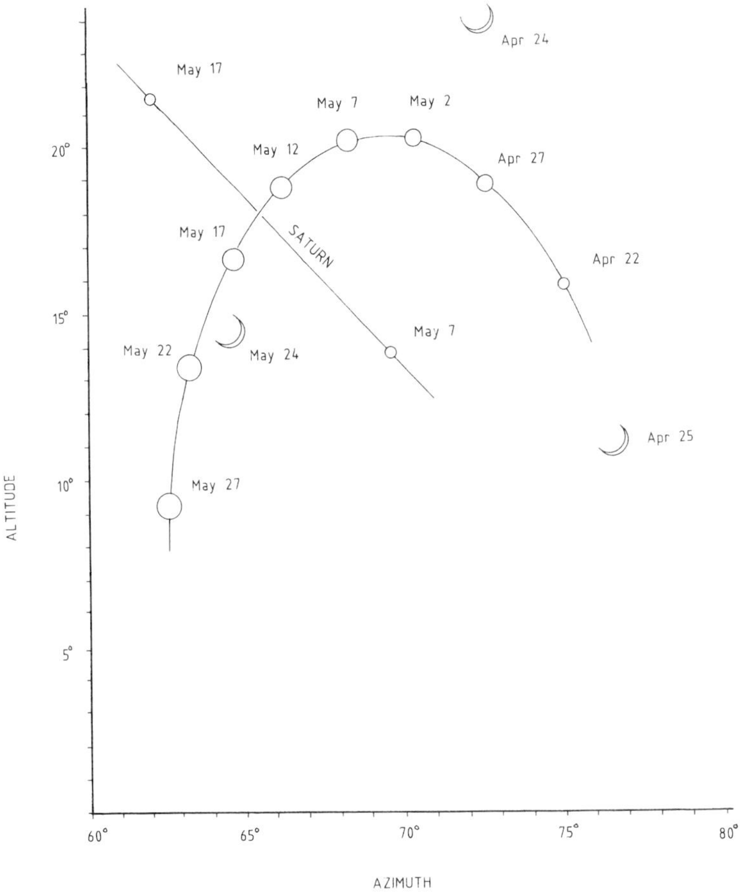

Figure 4. Morning apparition of Mercury, from latitude 35°S.

morning twilight. For those further south, Saturn very gradually becomes visible during the second half of the month, low above the eastern horizon before the morning twilight inhibits observation. Its magnitude is +0.6.

JUNE

New Moon: June 24 *Full Moon*: June 10

Solstice: June 21

MERCURY passes through superior conjunction on June 10 and therefore will not be visible to observers in northern temperate latitudes during the month. Observers nearer the equator should be able to detect the planet during the last week of June, low in the west-north-western sky at the end of evening civil twilight. Readers should refer to Figure 5, given with the notes for July.

VENUS is a splendid object in the morning skies, at magnitude −3.9, visible above the eastern horizon before dawn. Observers in the British Isles will find that it is visible for a little longer each morning as the month progresses. This effect is caused by its northward movement in declination which more than offsets the fact that Venus is slowly moving in towards the Sun.

MARS is too close to the Sun for observation.

JUPITER, magnitude −2.4, continues to be visible as a morning object in the south-eastern sky. By the end of the month, observers at the latitudes of the British Isles will be able to see Jupiter shortly after midnight.

SATURN, magnitude +0.5, is a morning object in the eastern sky, though not visible to observers at the latitudes of the British Isles until nearly the end of June.

JULY

New Moon: July 23 *Full Moon*: July 9

EARTH is at aphelion (furthest from the Sun) on July 4, at a distance of 152 million kilometres.

MERCURY continues to be visible as an evening object, though not to observers in northern temperate latitudes, due to the long twilight and the fact that Mercury is further south in declination

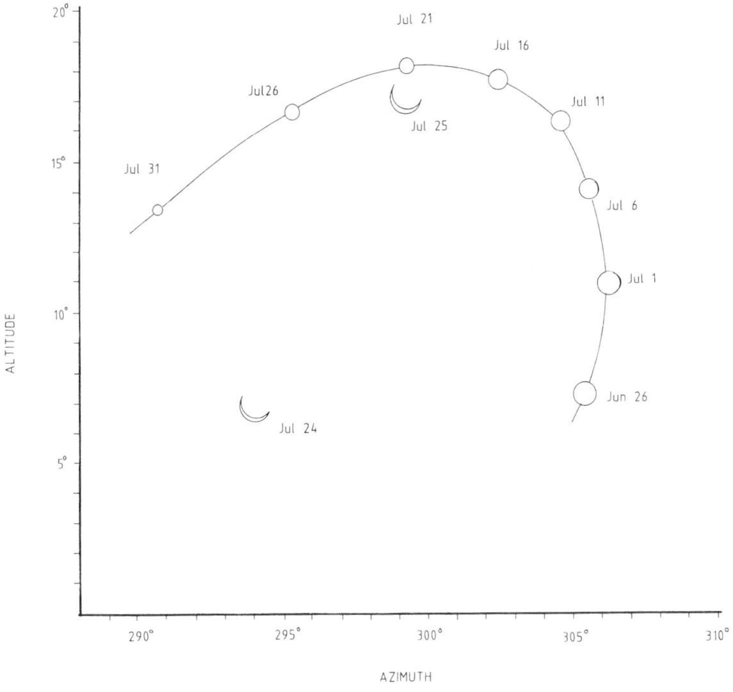

Figure 5. Evening apparition of Mercury, from latitude 35°S.

than the Sun. For observers in southern latitudes this will be the most favourable evening apparition of the year. Figure 5 shows, for observers at latitude 35°S, the changes in azimuth (the true bearing from the north through east, south and west) and altitude of Mercury on successive evenings when the Sun is 6° below the horizon. This condition is known as the end of evening civil twilight, and for this latitude and time of year occurs about 30 minutes after sunset. The changes in the brightness of the planet are indicated by the relative sizes of the circles marking Mercury's position at five-day intervals. Mercury is at its brightest before it reaches greatest eastern elongation (27°) on July 17.

VENUS, magnitude −3.9, continues to be visible as a brilliant object in the eastern morning skies before sunrise. From the latitudes of the British Isles the planet is visible low in the east-north-east before 03^h. On July 3, Venus passes 4° north of Aldebaran.

JUPITER is a morning object in the south-eastern sky, rising just south of east in the late evening. Its magnitude is −2.6.

SATURN is a morning object in the eastern sky, magnitude +0.5. For observers in the British Isles it is visible low above the eastern horizon shortly after 23^h by the end of July.

NEPTUNE is at opposition on July 23, in the eastern part of Sagittarius. It is not visible to the naked eye since its magnitude is +7.8. At opposition Neptune is 4357 million kilometres from Earth.

AUGUST

New Moon: August 22 *Full Moon*: August 8

MERCURY passes through inferior conjunction on August 13 and is thus too close to the Sun for observation. At the very end of the month it becomes visible as a morning object low in the eastern sky about half an hour before sunrise for observers at the latitudes of the British Isles. For observers at northern temperate latitudes, this

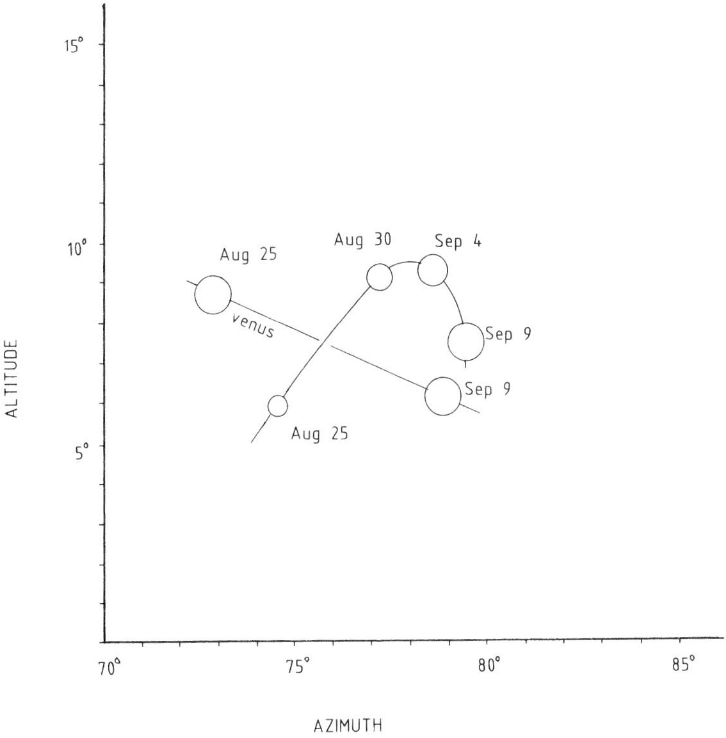

Figure 6. Morning apparition of Mercury, from latitude 52°N.

will be the most favourable morning apparition of the year. Figure 6 shows, for observers at latitude 52°N, the changes in azimuth (the true bearing from the north through east, south and west) and altitude of Mercury on successive mornings when the Sun is 6° below the horizon. This condition is known as the beginning of morning civil twilight, and for this latitude and time of year occurs about 35 minutes before sunrise. The changes in the brightness of the planet are indicated by the relative sizes of the circles marking Mercury's position at five-day intervals. Mercury is at its brightest after it reaches greatest western elongation (18°) on August 31. Venus, four magnitudes brighter, will be a useful guide to locating Mercury, passing 2° north of it on August 25.

VENUS continues to be visible as a splendid morning object, magnitude −3.9. It is clearly seen above the east-north-eastern horizon before sunrise.

MARS, magnitude +1.7, very gradually becomes a morning object during August, visible low above the east-north-eastern horizon before the morning twilight inhibits observation. Venus could be used as a guide during the first half of the month, when the two planets are quite close; on the mornings of August 4 and 5 the separation is less than 1°.

JUPITER, magnitude −2.8, continues to be visible as a brilliant object in the night sky. By the end of August it is visible low in the eastern sky by 20^h.

SATURN, magnitude +0.3, continues to be visible as a morning object, though now rising over the eastern horizon in the late evening.

URANUS is at opposition on August 3, in Capricornus. The planet is only just visible to the naked eye under the best of conditions since its magnitude is +5.7. In a small telescope it appears as a slightly greenish disk. At opposition Uranus is 2821 million kilometres from Earth.

SEPTEMBER

New Moon: September 20 *Full Moon*: September 6

Equinox: September 23

MERCURY, for observers at northern temperate latitudes, continues to be visible as a morning object for about the first fortnight of the month; reference should be made to Figure 6, given with the notes for August. Venus continues to be in the vicinity of Mercury, and passes only 0°.3 south of it on September 11.

VENUS, magnitude −3.9, remains a brilliant object low above the eastern horizon before dawn. However, it is getting noticeably closer to the Sun. By the end of September observers at the latitudes of the British Isles are likely to be able to see it only for about ten minutes or so before it is lost in the glare of the rising Sun.

MARS, magnitude +1.6, is a morning object, now visible in the eastern sky for several hours before dawn. Figure 7 shows its path amongst the stars during the rest of the year.

JUPITER, magnitude −2.9, is now observable throughout the hours of darkness as it passes through opposition on August 16. It is then 593 million kilometres from Earth. The path of Jupiter amongst the stars is shown in Figure 1, given with the notes for January.

SATURN is still a morning object, magnitude +0.1. Its path amongst the stars during the year is shown in Figure 2, given with the notes for February.

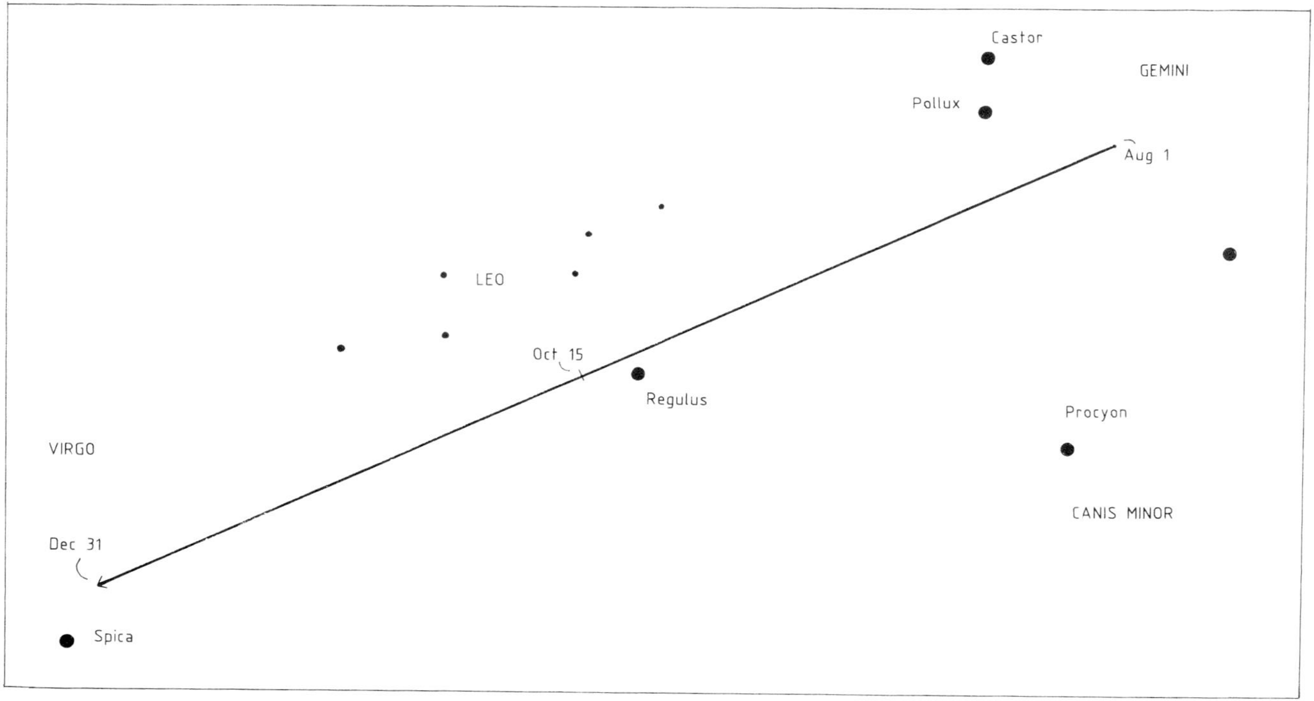

Figure 7. The path of Mars in 1998.

OCTOBER

New Moon: October 20 *Full Moon*: October 5

Summer Time in Great Britain and Northern Ireland ends on October 25.

MERCURY, for observers at equatorial and southern latitudes, gradually emerges from the evening twilight by the middle of the month, visible in the west-south-western sky at the end of evening civil twilight. Its magnitude is about −0.5. For those in the British Isles, Mercury is not suitably placed for observation.

VENUS is still a splendid morning object, magnitude −3.9, but only very low above the eastern horizon shortly before dawn. After the first week of the month it is too close to the Sun to be seen by those at northern latitudes. Observers further south are unlikely to glimpse the planet at all. Venus passes through superior conjunction on October 30.

MARS continues to be visible as a morning object, magnitude +1.5. It is now visible in the eastern sky for several hours before the morning twilight inhibits observation. Mars passes 1° north of Regulus on October 6, and is only very slightly fainter than the star.

JUPITER, just past opposition, is still a brilliant object in the night sky, magnitude −2.8. Jupiter is on the borders of Aquarius and Pisces.

SATURN, magnitude 0.0, reaches opposition on October 23 and thus remains visible throughout the hours of darkness. The rings are continuing to open up after the last passage of the Earth through the ring plane (in 1995). The minor axis of the rings is 12 arc seconds, while the polar diameter is 18 arc seconds. At opposition Saturn is 1241 million kilometres from Earth.

NOVEMBER

New Moon: November 19 *Full Moon*: November 4

MERCURY is at greatest eastern elongation (23°) on November 11 and therefore visible in the west-south-western sky around the time of end of evening civil twilight to observers at equatorial and southern latitudes (except for the last ten days of the month). During this period its magnitude fades from −0.2. to +1.0. However, for observers at the latitudes of the British Isles, Mercury remains unsuitably placed for observation.

VENUS is moving very slowly eastwards from the Sun, but even by the end of November it is still only 7° from the Sun, too close for observation.

MARS is still a morning object in the eastern sky, magnitude +1.2. During November Mars moves from Leo into Virgo (see Figure 7, given with the notes for September). The slight reddish tinge of the planet should assist in identification.

JUPITER, magnitude −2.6, continues to be visible as a brilliant object in the evening sky, but by the end of the month is no longer visible after midnight. Observers with a good pair of binoculars, provided they are steadily supported, should attempt to detect the four Galilean satellites. The main difficulty in observing them is the overpowering brightness of Jupiter itself.

SATURN, just past opposition, continues to be visible as an evening object, though actually above the horizon for the greater part of the night. Its magnitude is +0.1.

DECEMBER

New Moon: December 18 *Full Moon*: December 3

Solstice: December 22

MERCURY, for observers at northern temperate latitudes, emerges from the morning twilight after the first ten days of the month, passing through greatest western elongation (22°) on December 20. It may be seen low above the south-eastern horizon around the time of beginning of morning civil twilight. For observers at southern latitudes the period of visibility is restricted to the second half of the month. During its period of visibility its magnitude brightens from +1.2 to −0.4.

VENUS is too close to the Sun for observation at first, but gradually moves outwards and so becomes visible in the evening skies, magnitude −3.9, low above the south-western horizon, but only for a short while after sunset. Observers at northern temperate latitudes will have difficulty in seeing the planet at all until about ten days before the end of the year, because Venus is so far south of the celestial equator.

MARS continues to be visible as a morning object in the south-eastern sky. It is now rising over the eastern horizon before 02^h, and its magnitude brightens during December from +1.4 to +1.0.

JUPITER is still a conspicuous evening object, magnitude −2.4. By the end of December observers in the British Isles will find it crossing the meridian about an hour after sunset and being lost to view over the western horizon by 22^h.

SATURN, magnitude +0.3, is still visible as an evening object. It ends the year on the border of Aquarius and Pisces.

Eclipses in 1998

During 1998 there will be two eclipses of the Sun, but none of the Moon.

1. *A total eclipse of the Sun on February 26* is visible as a partial eclipse from the eastern Pacific Ocean, southern and eastern North America, central America, the northern part of South America, the North Atlantic Ocean, and the extreme western portions of the Iberian Peninsula and west Africa. It is just possible that observers could witness the beginning of the eclipse from extreme south-west Ireland, as it begins there at about 18^h 09^m, only a few minutes before sunset. The partial phase begins at 14^h 50^m and ends at 20^h 06^m. The track of totality starts in mid-Pacific and crosses extreme southern Panama, northern Colombia, and north-western Venezuela, passes through the Leeward Islands, and ends in the eastern North Atlantic Ocean. The total phase begins at 15^h 47^m and ends at 19^h 10^m; the maximum duration is 4^m 09^s.

2. *An annular eclipse of the Sun on August 21–22* is visible as a partial eclipse from eastern India and the other countries of South-east Asia, Indonesia, Philippines, Australasia, and western and southern parts of the Pacific Ocean. It begins at 21^d 23^h 10^m and ends at 22^d 05^h 02^m. The annular phase commences in the eastern Indian Ocean, crosses northern Sumatra, Malaysia, northern Borneo, New Britain, New Hebrides and Vanuatu, and ends in the South Pacific Ocean. The annular phase begins at 22^d 00^h 14^m and ends at 22^d 03^h 58^m; the maximum duration is 3^m 14^s.

Occultations in 1998

In the course of its journey round the sky each month, the Moon passes in front of all the stars in its path, and the timing of these occultations is useful in fixing the position and motion of the Moon. The Moon's orbit is tilted at more than 5° to the ecliptic, but it is not fixed in space. It twists steadily westwards at a rate of about 20° a year, a complete revolution taking 18.6 years, during which time all the stars that lie within about 6½° of the ecliptic will be occulted. The occultations of any one star continue month after month until the Moon's path has twisted away from the star, but only a few of these occultations will be visible from any one place in hours of darkness.

There are twenty-three lunar occultations of bright planets in 1998: two of Mercury, three of Venus, three of Mars, eleven of Jupiter, and four of Saturn.

Only four first-magnitude stars are near enough to the ecliptic to be occulted by the Moon: Aldebaran, Regulus, Spica and Antares. Aldebaran undergoes occultation thirteen times and Regulus eight times in 1998.

Predictions of these occultations are made on a worldwide basis for all stars down to magnitude 7.5, and sometimes even fainter. The British Astronomical Association has produced a complete lunar occultation prediction package for microcomputer users.

Recently, occultations of stars by planets (including minor planets) and satellites have aroused considerable attention.

The exact timing of such events gives valuable information about positions, sizes, orbits, atmospheres and sometimes of the presence of satellites. The discovery of the rings of Uranus in 1977 was the unexpected result of the observations made of a predicted occultation of a faint star by Uranus. The duration of an occultation by a satellite or minor planet is quite small (usually of the order of a minute or less). If observations are made from a number of stations it is possible to deduce the size of the planet.

The observations need to be made either photoelectrically or visually. The high accuracy of the method can readily be appreciated when one realizes that even a stop-watch timing accurate to $0^s.1$ is, on average, equivalent to an accuracy of about 1 kilometre in the chord measured across the minor planet.

Comets in 1998

The appearance of a bright comet is a rare event which can never be predicted in advance, because this class of object travels round the Sun in enormous orbits with periods which may well be many thousands of years. There are therefore no records of the previous appearances of these bodies, and we are unable to follow their wanderings through space.

Comets of short period, on the other hand, return at regular intervals, and attract a good deal of attention from astronomers. Unfortunately they are all faint objects, and are recovered and followed by photographic methods using large telescopes. Most of these short-period comets travel in orbits of small inclination which reach out to the orbit of Jupiter, and it is this planet which is mainly responsible for the severe perturbations which many of these comets undergo. Unlike the planets, comets may be seen in any part of the sky, but since their distances from the Earth are similar to those of the planets their apparent movements in the sky are also somewhat similar, and some of them may be followed for long periods of time.

The following periodic comets are expected to return to perihelion in 1998, and to be brighter than magnitude +15:

Comet	*Year of discovery*	*Period (years)*	*Predicted date of perihelion 1998*
Tempel–Tuttle	1865	33.2	Feb. 27
Howell	1981	5.6	Sep. 27
Lovas 1	1980	9.1	Nov. 3
Giacobini–Zinner	1900	6.6	Nov. 21

Minor Planets in 1998

Although many thousands of minor planets (asteroids) are known to exist, only a few thousand of these have well-determined orbits and are listed in the catalogues. Most of these orbits lie entirely between the orbits of Mars and Jupiter. All of these bodies are quite small, and even the largest, Ceres, is only 913 km (567 miles) in diameter. Thus, they are necessarily faint objects, and although a number of them are within the reach of a small telescope few of them ever reach any considerable brightness. The first four that were discovered are named Ceres, Pallas, Juno and Vesta. Actually the largest four minor planets are Ceres, Pallas, Vesta and Hygeia (excluding 2060 Chiron, which orbits mainly between the paths of Saturn and Uranus, and whose nature is uncertain). Vesta can occasionally be seen with the naked eye and this is most likely to occur when an opposition occurs near June, since Vesta would then be at perihelion. Ephemerides for these minor planets in 1998 are:

1 Ceres

		2000.0 RA		Dec.		*Geo-centric distance*	*Helio-centric distance*	*Phase angle*	*Visual magni-tude*	*Elonga-tion*
		h	*m*	°	′	*AU*	*AU*	°		°
July	6	3	27.21	+12	20.7	3.376	2.855	16.2	9.2	51.5W
	16	3	40.58	+13	10.5	3.261	2.848	17.5	9.1	57.6W
	26	3	53.40	+13	54.1	3.138	2.841	18.7	9.1	63.8W
Aug.	5	4	5.51	+14	31.4	3.009	2.834	19.7	9.0	70.3W
	15	4	16.73	+15	2.9	2.874	2.826	20.4	8.9	77.1W
	25	4	26.85	+15	28.8	2.735	2.818	20.9	8.8	84.2W
Sep.	4	4	35.62	+15	49.8	2.594	2.811	21.0	8.7	91.7W
	14	4	42.76	+16	6.6	2.454	2.803	20.7	8.6	99.6W
	24	4	47.97	+16	20.0	2.317	2.795	20.0	8.4	108.0W
Oct.	4	4	50.93	+16	30.9	2.186	2.787	18.6	8.3	117.1W
	14	4	51.36	+16	40.3	2.065	2.779	16.7	8.1	126.7W
	24	4	49.07	+16	49.1	1.958	2.771	14.1	7.8	137.1W
Nov.	3	4	44.06	+16	58.0	1.870	2.763	10.9	7.6	148.2W
	13	4	36.63	+17	7.4	1.805	2.755	7.1	7.4	159.8W
	23	4	27.38	+17	17.8	1.766	2.747	3.1	7.1	171.4W

1 Ceres – *cont.*

Dec.	3	4	17.33	+17	29.8	1.757	2.739	2.3	7.0	173.7E
	13	4	7.61	+17	44.4	1.776	2.731	6.3	7.3	162.3E
	23	3	59.30	+18	2.8	1.824	2.723	10.3	7.5	150.4E

2 Pallas

		2000.0								
		RA		*Dec.*		*Geo-centric distance*	*Helio-centric distance*	*Phase angle*	*Visual magni-tude*	*Elonga-tion*
		h	*m*	°	′	*AU*	*AU*	°		°
July	6	23	54.58	+7	37.6	2.829	3.197	18.1	9.9	101.9W
	16	23	57.45	+7	13.3	2.675	3.182	17.4	9.7	110.8W
	26	23	58.53	+6	30.5	2.528	3.166	16.1	9.6	120.4W
Aug.	5	23	57.70	+5	27.0	2.394	3.149	14.2	9.4	130.6W
	15	23	54.90	+4	1.5	2.276	3.132	11.6	9.1	141.4W
	25	23	50.24	+2	14.4	2.181	3.115	8.5	8.9	152.8W
Sep.	4	23	44.02	+0	8.7	2.112	3.097	4.9	8.6	164.8W
	14	23	36.78	−2	9.6	2.073	3.078	0.9	8.3	177.2W
	24	23	29.22	−4	32.0	2.066	3.059	3.2	8.5	170.3E
Oct.	4	23	22.12	−6	49.1	2.090	3.040	7.1	8.7	157.9E
	14	23	16.24	−8	52.5	2.142	3.020	10.7	8.9	145.8E
	24	23	12.11	−10	36.5	2.220	3.000	13.7	9.1	134.2E
Nov.	3	23	10.11	−11	58.1	2.317	2.979	16.2	9.2	123.3E
	13	23	10.34	−12	57.3	2.429	2.958	17.9	9.4	112.9E
	23	23	12.77	−13	35.2	2.549	2.937	19.1	9.5	103.2E
Dec.	3	23	17.27	−13	53.9	2.675	2.915	19.7	9.6	94.0E
	13	23	23.60	−13	56.2	2.800	2.893	19.8	9.7	85.4E
	23	23	31.56	−13	44.2	2.922	2.870	19.5	9.8	77.2E

3 Juno

		2000.0								
		RA		*Dec.*		*Geo-centric distance*	*Helio-centric distance*	*Phase angle*	*Visual magni-tude*	*Elonga-tion*
		h	*m*	°	′	*AU*	*AU*	°		°
Jan.	27	12	24.20	−3	6.7	2.109	2.739	18.1	9.9	120.1W
Feb.	6	12	23.58	−2	23.1	2.019	2.765	15.7	9.7	130.7W
	16	12	20.54	−1	20.6	1.946	2.791	12.6	9.6	141.9W
	26	12	15.31	−0	1.7	1.894	2.816	9.0	9.4	153.6W
Mar.	8	12	8.37	+1	28.8	1.869	2.841	4.9	9.3	165.7W
	18	12	.52	+3	3.6	1.871	2.865	1.2	9.1	176.7W
	28	11	52.62	+4	34.6	1.904	2.889	3.8	9.3	168.8E
Apr.	7	11	45.55	+5	54.7	1.965	2.913	7.7	9.6	157.0E
	17	11	40.02	+6	58.6	2.053	2.936	11.1	9.8	145.6E

4 Vesta

	RA 2000.0 h	m	Dec. °	′	Geo-centric distance AU	Helio-centric distance AU	Phase angle °	Visual magni-tude	Elonga-tion °
Jan. 7	1	26.73	+1	37.0	2.272	2.539	22.7	7.9	94.1E
17	1	34.69	+2	59.6	2.410	2.544	22.7	8.0	86.4E
27	1	44.22	+4	26.9	2.547	2.549	22.3	8.1	79.0E
Feb. 6	1	55.11	+5	57.2	2.680	2.553	21.5	8.2	71.9E
16	2	7.16	+7	28.9	2.809	2.557	20.5	8.3	65.2E
26	2	20.20	+9	0.6	2.931	2.560	19.3	8.4	58.7E
Mar. 8	2	34.13	+10	31.1	3.045	2.563	17.9	8.4	52.4E
18	2	48.81	+11	59.1	3.150	2.566	16.3	8.4	46.3E
28	3	4.17	+13	23.8	3.245	2.568	14.6	8.5	40.4E
Apr. 7	3	20.14	+14	44.1	3.330	2.570	12.8	8.5	34.7E
Aug. 15	7	13.08	+21	23.9	3.321	2.559	13.1	8.5	35.0W
25	7	30.50	+21	1.8	3.236	2.555	14.9	8.4	40.6W
Sep. 4	7	47.44	+20	34.4	3.142	2.551	16.6	8.4	46.3W
14	8	3.82	+20	2.9	3.037	2.546	18.2	8.4	52.1W
24	8	19.54	+19	28.5	2.924	2.542	19.6	8.3	58.1W
Oct. 4	8	34.47	+18	52.6	2.803	2.537	20.8	8.3	64.4W
14	8	48.49	+18	17.1	2.675	2.531	21.9	8.2	70.9W
24	9	1.45	+17	43.6	2.542	2.525	22.6	8.1	77.7W
Nov. 3	9	13.16	+17	14.6	2.405	2.519	23.1	8.0	84.9W
13	9	23.40	+16	52.3	2.267	2.513	23.2	7.9	92.5W
23	9	31.92	+16	39.5	2.129	2.506	22.8	7.7	100.6W
Dec. 3	9	38.40	+16	38.8	1.995	2.499	21.9	7.5	109.2W
13	9	42.52	+16	52.7	1.867	2.491	20.3	7.3	118.5W
23	9	43.95	+17	23.4	1.749	2.483	18.1	7.1	128.4W

A vigorous campaign for observing the occultations of stars by the minor planets has produced improved values for the dimensions of some of them, as well as the suggestion that some of these planets may be accompanied by satellites. Many of these observations have been made photoelcctrically. However, amateur observers have found renewed interest in the minor planets since it has been shown that their visual timings of an occultation of a star by a minor planet are accurate enough to lead to reliable determinations of diameter. As a consequence many groups of observers all over the world are now organizing themselves for expeditions should the predicted track of such an occultation cross their country.

In 1984 the British Astronomical Association formed a special Asteroid and Remote Planets Section.

Meteors in 1998

Meteors ('shooting stars') may be seen on any clear moonless night, but on certain nights of the year their number increases noticeably. This occurs when the Earth chances to intersect a concentration of meteoric dust moving in an orbit around the Sun. If the dust is well spread out in space, the resulting shower of meteors may last for several days. The word 'shower' must not be misinterpreted – only on very rare occasions have the meteors been so numerous as to resemble snowflakes falling.

If the meteor tracks are marked on a star map and traced backwards, a number of them will be found to intersect in a point (or a small area of the sky) which marks the radiant of the shower. This gives the direction from which the meteors have come.

The following table gives some of the more easily observed showers with their radiants; interference by moonlight is shown by the letter M.

Limiting dates	*Shower*	*Maximum*	*RA*		*Dec.*	
			h	*m*	°	
Jan. 1–4	Quadrantids	Jan. 3	15	28	+50	
Apr. 20–22	Lyrids	Apr. 22	18	08	+32	
May 1–8	Eta Aquarids	May 4	22	20	−01	
June 17–26	Ophiuchids	June 19	17	20	−20	
July 15–Aug. 15	Delta Aquarids	July 29	22	36	−17	
July 15–Aug. 20	Piscis Australids	July 31	22	40	−30	
July 15–Aug. 25	Capricornids	Aug. 2	20	36	−10	
July 27–Aug. 17	Perseids	Aug. 12	3	04	+58	M
Oct. 15–25	Orionids	Oct. 21	6	24	+15	
Oct. 26–Nov. 16	Taurids	Nov. 3	3	44	+14	M
Nov. 15–19	Leonids	Nov. 17	10	08	+22	
Dec. 9–14	Geminids	Dec. 13	7	28	+32	
Dec. 17–24	Ursids	Dec. 23	14	28	+78	

Some Events in 1999

ECLIPSES

There will be three eclipses, two of the Sun and one of the Moon.

February 16:	annular eclipse of the Sun – southern Africa, Australasia
July 28:	partial eclipse of the Moon – part of the Americas, Australasia
August 11:	total eclipse of the Sun – north-eastern North America, Europe, north Africa, Asia.

THE PLANETS

Mercury may be seen more easily from northern latitudes in the evenings about the time of greatest eastern elongation (March 3) and in the mornings around greatest western elongation (August 14). In the Southern Hemisphere the corresponding most favourable dates are around April 16 (mornings) and October 24 (evenings).

Venus is visible in the evenings from the beginning of the year until mid-August. From late August until the end of the year it is visible in the mornings.

Mars is at opposition on April 24.

Jupiter is at opposition on October 23.

Saturn is at opposition on November 6.

Uranus is at opposition on August 7.

Neptune is at opposition on July 26.

Pluto is at opposition on May 31.

Astronomers – and squirrels

PATRICK MOORE

To astronomers, seeing conditions are all-important. To have a good telescope, and use it under poor conditions, is tantamount to having a good record-player and a bad stylus. Mountain-tops are particularly suitable, because there is less overlying atmosphere to cause distortion; this is why high-altitude sites, such as the summit of Mauna Kea in Hawaii and the top of Roque de los Muchachos in the Canary Islands, are now major astronomical centres. This sort of consideration was very much in the minds of the astronomers at the Vatican Observatory when they started to think about a new location.

The Vatican Observatory has a long and honourable history, but it would be idle to pretend that conditions at the original site are good. A move had been made well out of Rome, but even this was not really satisfactory. There was no need to stay in Europe. Various opportunities were offered in Arizona, and Mount Graham, in the Pinaleños Mountains, seemed to be an excellent choice. It is lofty; the air is stable, and there is no light pollution of any significance. So Mount Graham was selected.

Two main telescopes were planned, and in fact both are now in operation. One is at the Submillimeter Telescope Observatory, and is officially called the Heinrich Hertz Telescope in honour of the German scientist who first demonstrated the electromagnetic nature of light; this telescope was the result of a collaboration initially between the Max Planck Institute in Germany itself and the University of Arizona. It is a 10-metre telescope, designed specifically for observations at wavelengths between 0.3 and 1.3 mm – ideal for studying objects such as the cool, dusty molecular clouds inside which stars are being formed. The other was the VATT, or Vatican Advanced Technology Telescope, named officially the Alice P. Lennon Telescope – a prototype for a new generation of optical telescopes; the mirror is 1.8 metres in diameter, and is very 'fast', with a focal ratio of 1 (*f*/1). This design means that the

Figure 1. The Heinrich Hertz Telescope at the Submillimeter Telescope Observatory.

Figure 2. The Alice P. Lennon Telescope.

Figure 3. The dome of the Vatican Advanced Technology Telescope (VATT).

telescope is short and squat, so that the observatory for it can be very compact. The mounting is of the altazimuth type, and of course the telescope is completely computerized. The mirror was made by Roger Angell, and is of excellent quality. It is a very versatile telescope, used for optical and near-infrared imaging, optical astronomy and polarimetry. The programmes include studies of the magnetic fields in regions of star formation, and searches for the optical and infrared counterparts of radio and X-ray sources.

All seemed well, but the astronomers had reckoned without one complication: squirrels.

The Pinaleños are made up of a series of 'mountain archipelagos', in which the summits are to some extent isolated from their surroundings. It so happened that Mount Graham was the habitat of some very rare red squirrels. There are not very many of them (a few hundred), but they are strictly protected by law – even though it is only a few years ago that they were targets for hunters with guns! It is quite true that they qualify as 'endangered', and as soon as the observatory plans became public the conservationists were up in arms. The astronomers, they claimed, would destroy much of the squirrels' habitat, and could well cause their extinction in the area.

The objection became really vehement. 'No Scopes' stickers appeared in the campus of the University of Tucson; protests grew, and were backed by a radical group calling itself 'Earth First!'. One Arizona biologist who supported the telescope scheme even received death threats by mail. More than a dozen environmental groups joined in, claiming that the astronomers had cut across the national Environmental Policy Act and the Endangered Species Act by obtaining a special exemption from Congress.

Battle was joined in the early 1980s. But what were the real facts?

First, had there been any real danger of the squirrels being wiped out, it is very likely that the astronomers would have accepted the situation and gone elsewhere. But it did not seem that there was any real possibility of reducing the squirrels' numbers: the habitat is admittedly restricted, but the observatory would take very little space, and in any case the squirrels can function quite well in adjacent areas. Exhaustive studies were carried out by all sorts of organizations, both political and conservationist, and finally, in the summer of 1990, permission was granted. The authorities lost no time in starting to clear a suitable piece of ground.

Everything was done to avoid causing alarm. The site of the observatory is indeed small; round it are fences which are strictly 'out of bounds', and the squirrels are left undisturbed. The ironical fact is that, after observatory building started, the numbers of squirrels increased rather than declined. Moreover, the observatory itself is host to biological teams studying not only the red squirrels, but also other important ecological features of the site.

It cannot be said that the battle is over, but there is at least a truce which must surely develop into a permanent peace. So the telescopes on Mount Graham can continue with their work unhindered – and astronomers and squirrels exist side by side in harmony. It is very different from the bitter battles of the 1980s, when for a while the astronomers were entitled to feel that they were being driven nuts!

Life on Mars?

RICHARD L. S. TAYLOR

Have We Found Life on Mars?

Perhaps the single most important discovery we could make would be to find incontrovertible evidence for the existence of life elsewhere in the Universe. One thing is certain: the confirmation of the existence of extraterrestrial life will change for ever the way we see ourselves and our relationship to the Universe. This makes it important that we do not rush to premature conclusions on what may appear to be traces of possible past biological activity within meteorites, whether from Mars or elsewhere – particularly when those traces lie close to the limits of detection of our current instruments and techniques of measurement. Given these considerations, what are we to make of the claim that traces of ancient life may exist in a rock from Mars?

On August 7, 1996, NASA held a press conference to announce that a research team led by David McKay and Everett Gibson had found traces of biogenic materials, and what appeared to be fossilized remains of ancient micro-organisms, within ALH 84001, a meteorite identified as having come from Mars. Excitement in the popular press and news media rose rapidly to fever pitch, and the claimed discovery of extraterrestrial life was treated more as a certainty than a possibility. Scientists were more cautious in evaluating the evidence presented, partly because supposed discoveries of extraterrestrial organic compounds, complex molecules and even microbe-like objects in meteorites have a long and rather chequered history.

The Swedish chemist Jöns Jacob Berzelius was perhaps the first to announce that he had extracted a whole complex of organic compounds, from the 1834 Alais meteorite. Friedrich Wöhler made similar claims for the meteorite that fell at Kaba, Hungary, in 1857. A more problematic case was that of the 1864 Orgueil meteorite. Four years later the French scientist Marcellin Berthelot claimed to have extracted saturated hydrocarbons from it, and went so far as to compare them to the hydrocarbons in petroleum. In the early 1960s, B. Nagy and G. Claus re-examined the Orgueil meteorite and reported detecting a variety of organic compounds, and also organized elements – microfossils – within it. This claim did not last long. It soon transpired that the organized elements in the meteorite were mineral and pollen grains, the latter from terrestrial contamination, and not evidence for extraterrestrial life. The 1969 Murchison fall in Victoria, Australia, is a more recent example of a meteorite confirmed to contain organic molecules. This meteorite exploded just before impact, scattering débris over a wide area. About 5 kg of fragments were recovered, and samples analysed at the NASA Ames Laboratories were found to contain five of the twenty-three amino acids common to all living cells, together with lesser amounts of eleven other protein-forming amino acids.

Organic molecules exhibit the property of chirality; that is, they exist in right- and left-handed (dextro and laevo) forms. In biological materials the molecules are exclusively of the left-handed form, but in the Murchison meteorite the dextro and laevo molecular forms were present in equal quantities. This condition, termed *racemization*, is characteristic of a non-biological origin. The important point here is that the mere presence of amino acids in extraterrestrial material is not in itself a sure indicator of past biological activity. In the case of ALH 84001 it is the distribution and relationship of the organic compounds that constitutes evidence for their endogenous origin, and their association with specific structural microbe-like elements that appears consistent with a possible biogenic origin. But just how well founded is the NASA evidence that there may have been life on Mars?

ALH 84001 (Figure 1) is a 1.9-kg meteorite collected from the Allan Hills region of Antarctica in 1984, by NASA scientist Roberta Score, then a member of the Antarctic Meteorite Search Team. Because its mineral composition and crystallization age of around 4.5 billion years make ALH 84001 quite different from the other eleven meteorites currently believed to have come from Mars, it was only identified as being of Martian origin in 1993. The other eleven form the SNC[1] subgroup of the class of meteorites known as achondrites, and all have crystallization ages of around 1.4 billion years. Examination of ALH 84001 showed that it had fallen about 13,000 years ago, and its cosmic-ray exposure revealed that before colliding with the Earth it had spent some 16 million years in space. Following the identification of ALH 84001 as a rock from Mars, David McKay and his colleagues spent rather more than two years studying this unique Martian rock sample before announcing their discoveries and publishing them in the journal *Science*.[2]

Figure 1. The meteorite designated ALH 84001, recovered from Antarctica where it had lain for 13,000 years, having spent 16 million years in space after being blasted out of the surface of Mars by a low-angle asteroidal impact. The cube measures 1 cm.

Meteorites from Mars?

Did ALH 84001 and the SNC meteorites actually come from Mars? A Martian origin for the SNC meteorites was proposed in the early 1980s, when it was discovered that glassy nodules embedded in the basaltic rock of the meteorite known as EETA 79001 contained a noble gas component which is a detailed match for the same gases in the Martian atmosphere as measured by the Viking spacecraft. Other supporting evidence soon followed. The enrichment of the heavier isotope of nitrogen (^{15}N) in SNC meteorites also matched the Viking data for the Martian atmosphere, as did the ^{13}C:^{12}C carbon and ^{17}O:^{18}O oxygen isotope ratios. All of these are essentially the same for the SNC meteorite bodies, ALH 84001 and Mars. If we add to this a comparison of the mineral composition of SNC meteorites with the Martian material analysed by the Viking lander, we find other compelling similarities.

If analytical data suggests a Martian origin for SNC meteorites and ALH 84001, how did they get to Earth? The majority of all meteorites that reach the surface of the Earth remain unidentified and uncollected, so the twelve known Martian meteorites must represent only a very small fraction of the mass of material which, ejected from Mars, has eventually reached the Earth. (It has been estimated that 500 kg of Mars rock falls on Earth each year.) When an impact takes place on a planet, pieces of rock much broader than they are thick, called spall plates, are ejected from the surface. At ejection velocities greater than 0.5 km/s these plates contain so much elastic energy that they break up into many smaller pieces, and some of these solid fragments have velocities that may approach half the original impact velocity. At the orbital distance of Mars typical impact velocities are around twice the escape velocity of the planet, so it is possible for rocks to be ejected from Mars by impacts.

Once in space, how long does it take for a Martian rock fragment to reach Earth and fall as a meteorite? B. J. Gladman and J. A. Burns have used a computer simulation to examine the behaviour of rocks launched from Mars at speeds slightly above escape velocity. Taking account of close planetary encounters, and the effect of more distant planetary perturbations, the orbits of these ejected Martian rocks evolve over time, and after a few million years a small percentage of them collide with Earth. The 16-million-year transfer time for ALH 84001 is actually longer than average. An unexpected discovery was that about one meteorite fragment in ten million could strike Earth having spent less than half an orbital period – about one Earth year – in space. This raises the intriguing possibility that viable Martian micro-organisms might survive the rigours of space, and entry through Earth's atmosphere, and could have brought life to Earth from Mars, making us all Martians, the descendants of immigrant organisms!

So, the case for the SNC meteorites plus ALH 84001 having an origin on Mars is supported by several quite distinct strands of evidence. This does not entirely rule out the possibility that some could have come from another source, but the probability that these meteorites came from Mars is so high as to make it a near-certainty.

ALH 84001: Signs of Biological Activity Within

The evidence for possible ancient biological activity in ALH 84001 takes a number of forms. In material terms, ALH 84001 is an igneous rock with a mineral composition consisting chiefly of coarse-grained orthopyroxene, and a number of other minor constituent minerals. The bulk material of ALH 84001 crystallized about 4.5 billion years ago, making the rock almost as old as Mars itself. There is evidence of two subsequent shock events – one about 4 billion years ago, and the other only about 16 million years ago, around the time the meteorite was ejected from Mars.

Unlike the SNC meteorites, which contain only trace carbonates, ALH 84001 contains secondary carbonates that formed later than the main bulk of the rock. These take the form of globules about 1–250 μm across. It has been estimated by S. F. Knott and others that the globules formed about 3.6 billion years ago. But this age has been queried on the basis of strontium/rubidium ratio calculations by scientists of the Field Museum, Chicago. They suggest that the carbonates may be much younger, about 1.4 billion years old.

The mode of formation of the carbonate globules has also been a matter of dispute. Some researchers conclude that the carbonates formed at high temperature, between 450 and 700°C, too high for a biological origin. Others, on the basis of oxygen isotope data, conclude that the carbonates formed between 0 and 80°C, a temperature range entirely consistent with biological activity. However, if the preferential loss of the ^{16}O oxygen isotope from the Martian atmosphere is taken into account, this low temperature range has to be revised upwards to 40–250°C, a range only partially consistent with biological activity. At the present time the temperature problem has not been resolved, but the results obtained by different techniques appear to be converging on a range of values only partially consistent with biochemical activity, so that additional evidence, other than just that of temperature, may be necessary to decide between a biological or physico-chemical mode of origin.

An important constraint on the time when the carbonate globules formed comes from the fact that some of the globules were shock-faulted about 16 million years ago. This excludes the possibility that the globules formed in the rock while it was on Earth. McKay and his team have made very detailed studies of the textures of the carbonate globules, which appear to show a deposition/growth sequence that is common in terrestrial freshwater environments. On the basis of these observations, they propose that the globules are most likely the product of biological activity and were formed at low temperatures.

Polycyclic Aromatic Hydrocarbons

Kathie Thomas-Keprta, of NASA's Johnson Space Center, detected the presence of polycyclic aromatic hydrocarbons (PAHs) within ALH 84001. This is a group of organic compounds with complex ringed ('aromatic') structures. Certain members of the group are associated with the decay of terrestrial micro-organisms; others, some astronomers argue, may have their origin in giant molecular clouds in interstellar space. So the mere presence of PAHs in ALH 84001 would not constitute evidence for life if they were spread homogeneously through it, as would be expected had they been present at the time of the rock's formation. Richard Zare's unique microprobe two-step laser mass spectrometer enabled his team at Stanford University to confirm that the meteorite does contain abundant PAHs, and that these molecular species are not evenly distributed on the 50 μm scale. The average concentration of PAHs at interior fracture surfaces turns out to be greater than 1 part per million, a concentration at least a thousand times greater than expected in a meteorite. This concentration was also greatest in the carbonate-rich regions. McKay's team conclude that the presence of a complex mixture of organic molecules in ALH 84001 and their association with carbonates support the argument that they were formed through ancient biological activity.

One problem is whether the PAHs in the meteorite could be the result of terrestrial contamination. To try to answer this question, contamination checks and control experiments were performed by McKay's team. They found that the ALH 84001 PAHs are quite different from those expected from terrestrial contamination. PAHs in the Earth's atmosphere are primarily of industrial origin, and such pollution is very recent. Moreover, the measured concentrations of PAHs in the meteorite are between a thousand and a hundred thousand times greater than the maximum possible terrestrial PAHs levels on the Antarctic ice-cap.

Supporting evidence for the indigenous nature of ALH 84001's PAH content comes from studies of their distribution within the body of the meteorite. The exterior fusion crust and the zone immediately beneath it extending half a millimetre into the interior show a PAH signal that increases with depth. This concentration profile is consistent with the effects of volatilization and pyrolisis caused by the severe heating of the surface of the meteorite during atmospheric entry.

Another line of evidence for biological activity comes from the discovery of the coexistence of magnetite and iron sulphide phases within partially dissolved carbonate in ALH 84001. If these were formed non-biologically, a very complex explanation would be required to account for their presence. The magnetite particles are similar in chemistry, structure and form to common terrestrial magnetite particles known as magnetofossils – the remains of bacterial magnetosomes (typical size 20 to 100 nm). These form structures within a living cell which function as an orientation mechanism, a kind of compass; and after a cell dies, the magnetosomes survive. They are found in a variety of sedimentary deposits and soils on Earth, and are taken as clear evidence for past biological activity.

Given the lack of a strong magnetic field on Mars (its strength is only about one ten-thousandth of the Earth's), it is not clear what purpose magnetosome-like particles would serve in any Martian micro-organism. It is possible that the Martian magnetic field was significantly stronger in the past, but currently there is little to support such an assumption. These Martian magnetite particles may not have functioned as direction-finding aids within micro-organisms. At least some of the magnetite crystals in ALH 84001 resemble particles that are precipitated outside the cell during the growth of some strains of terrestrial bacteria. Such bacterial deposition of iron oxide particles is known to occur around deep-sea and surface hydrothermal vents on Earth.

Mars Microfossils: Evidence for Ancient Life?

The suggestion that what appear to be exceptionally small organized structures within ALH 84001 may be fossils of ancient Martian micro-organisms is a most exciting idea, but one that is subject to scepticism and doubt. The microstructures are of several different shapes and sizes, and occur within and on the carbonate surfaces, but even the largest of them are ten to a hundred times smaller than typical terrestrial bacteria (Figures 2 and 3). Their extremely small size has led McKay to compare them to microstructures found on the Earth in travertine and limestone, and interpreted by Bob Folk of the University of Texas as 'nanobacteria'. The evidence for nanobacteria is questioned by many scientists, who doubt that anything other than chemical or physical processes is needed to account for these minute structures within terrestrial rocks.

The extremely small size of these supposed Martian fossils is a significant problem, as there are sound theoretical reasons for

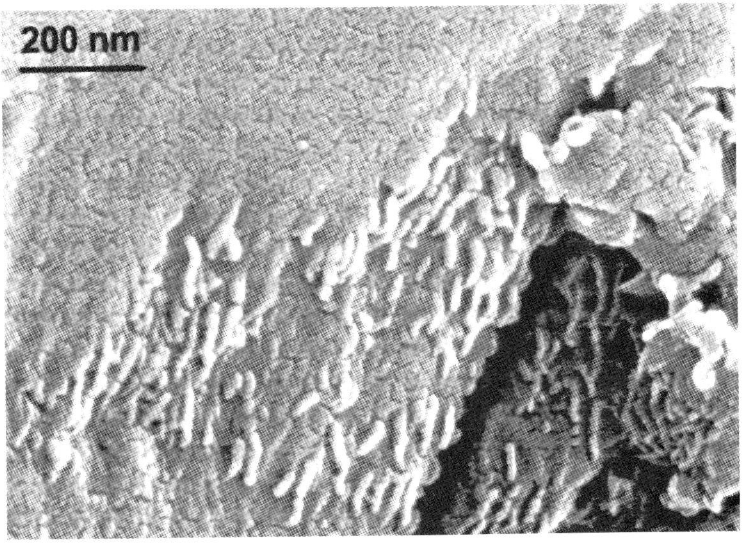

Figure 2. While the exact nature of these tube-like structures in ALH 84001 is not known, it is possible that they may be microscopic fossils of primitive bacterial organisms that lived on Mars more than 3.6 billion years ago. However, the largest structures visible here are no larger than the smallest known terrestrial virus, and most are ten times smaller.

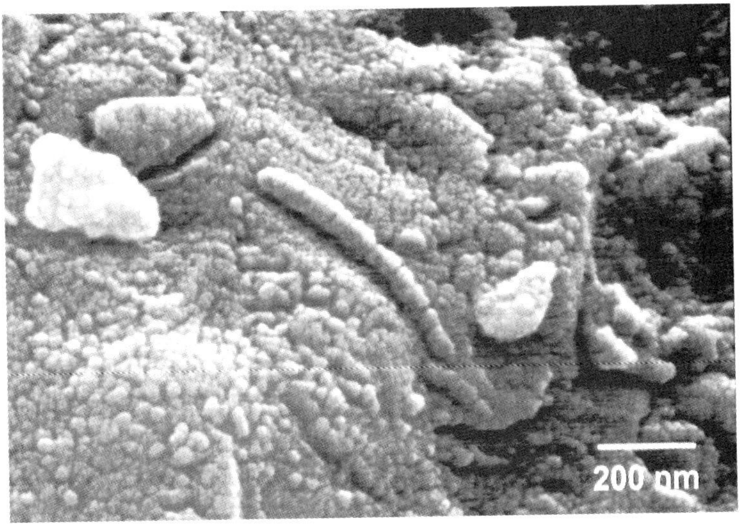

Figure 3. One of the most remarkable structures in ALH 84001 revealed by high-resolution microscopy is this tube-like object. Its length is only one-hundredth the diameter of a human hair – far smaller than the smallest viable terrestrial cellular organism.

believing there is a lower limit to the size of a free-living metabolizing cell. How critical is this size problem? Table 1 gives average dimensions for a range of terrestrial micro-organisms compared with those of the possible Martian micro-organisms. It can be seen that the dimensions of the Martian microstructures fall just above those of the smallest terrestrial viruses. They have only about 4% of the mass or volume of small terrestrial micro-organisms such as rickettsias, chlamydias or mycoplasmas, and only about 0.03% of the volume or mass of a typical medium-sized bacterium. Compared with a typical filiamentous cyanophyta (blue-green alga), the elongate form of the Martian structures is even more minute.

Terrestrial bacteria do not always retain their full size. During periods of long-term starvation many types of bacteria make use of their internal resources and shrink, decreasing from their normal size to about 10% of their original dimensions. In this shrunken state their volume may be reduced to ~0.1% of their normal values. T. L. Kieft, of the New Mexico Institute of Mining and Technology, has found by examining deep subsurface cores drilled from the Earth that such tiny, starved microbes commonly inhabit terrestrial rocks. The problem is whether a viral-sized entity like an ALH 84001 'fossil' can be anything but a subcellular non-metabolizing parasite, an organism capable only of living and replicating within a living cell.

Most terrestrial viruses consist of two constituents, protein and nucleic acid, the protein forming a viral coat, or capsid, around the nucleic acid. Exceptionally, a few viruses contain additional substances found in normal cells, for example the influenza virus contains also approximately 30% lipid and ~6% non-nucleic acid carbohydrate, and the vaccinia virus (smallpox) contains 6% lipid as well as nucleic acid and protein. Both these viruses are relatively large (influenza 200 × 80 nm, vaccinia 300 × 230 nm), and in having additional constituents they may be telling us something of their past evolution towards the viral state.

The rickettsias are some 30 to 1000 times larger in volume than the Martian fossils or terrestrial viruses. They were once classified as viruses, but are now considered to be close relations of bacteria. Their internal chemistry and organization are far more complex than those of a virus. Even so, of all the identified species of rickettsias only one is known that is not an obligatory parasite – in other words, like viruses almost all rickettsias require a living cell in which to grow and reproduce. However, many of the mycoplasmas,

Table 1. The lower limits to size for typical micro-organisms (in nanometres).

Object or organism	*Averaged diameter*	*Special characteristics*
Small molecules	0.45–0.85	—
Lipids	1.7–2.9	—
Proteins	3.1–10.0	—
Viruses and phages	40–100	Subcellular obligatory parasites. Contain either RNA or DNA plus protein coat. One or two species additionally contain small quantities of lipids or carbohydrates
Martian organisms, according to McKay *et al.*	~65 (Ovoid ~50, rod 50 × 100)	Both forms of 'microbe-like' structure are of viral rather than bacterial size. They are about 4% of the volume of the smallest examples of the rickettsias, chlamydias or mycoplasmas and less than 0.1% of the volume or mass of a typical bacterium
Mycoplasmas	125–460	No rigid cell wall present. Similar in form to the L-forms of true bacteria. Parasitic and saprophytic strains. Capable of growth on artificial media
Rickettsias	200–650	All but one species obligatory parasites. Classified as a form of bacteria
Chlamydias	200–1500	Obligatory parasites. Have a two-stage reproduction cycle involving two different structural forms
Normal bacteria	1000–10,000	Prokaryotic cells. Typical example *E. coli* is a rod ~2000 nm long, 800 nm diameter
Plant and animal cells	10,000–100,000	Eukaryotic cells containing organelles

organisms that are smaller than the rickettsias, exist in both parasitic and saprophytic (scavenging) strains. Mycoplasmas have no rigid cell wall and they contain less than half the quantity of nucleic acid present in any other fully functional cell. It is possible that this may represent the absolute minimum necessary for a metabolizing entity. Unlike viruses, chlamydias and most rickettsias, mycoplasmas are capable of growing on artificial culture media, but the parasitic strains require the culture medium to be enriched in proteins.

The lower limit to the size of a metabolizing cell is determined by the number of different types of molecule needed to perform the required range of biological functions. Such a cell contains some 30,000 different types of molecule. Living and fossil terrestrial organisms allow a crude estimate to be made of this lower size boundary. It appears that there is a transition zone covering the size range of the rickettsias and mycoplasmas, and that these two types of micro-organism have lost different functional elements during the process of evolutionary 'down-sizing'. The Mars micro-organisms are 10% of the mass and volume of the smallest mycoplasma, and thus appear to be an order of magnitude too small to have been, in life, fully functional metabolizing cells. This appears to rule out the minute ALH 84001 structures being the fossilized remains of free-living cellular Martian micro-organisms, although it is possible that we are actually looking at viral particles and have yet to find the micro-organisms they parasitize.

Fossils of terrestrial micro-organisms are mostly identifiable by a recognizable cell wall. Although the oldest examples of terrestrial micro-fossils are all of prokaryotic cellular organisms (cells with no distinct nucleus), it is sometimes possible to see trace evidence for the presence of complex internal membrane structures similar to those which occur in living prokaryotic organisms. However, all the well-established examples of terrestrial micro-fossils are at least a hundred times larger than the Martian examples, which in terms of size are more akin to the smallest structured organelles, or perhaps the nucleus, within a terrestrial eukaryotic cell (cells with a clearly defined nucleus). If McKay's micro-fossils are the remains of some form of organelle or nucleus with a fossilizable rigid cell-wall-like membrane, which in life was itself within a cell with a non-rigid wall (like a terrestrial mycoplasma), we can estimate probable upper and lower size limits of a living Martian micro-organism by extrapolating from what is known of the relative sizes of nuclei and organelles (e.g. mitochondria) in terrestrial cells. If the 100-nm Martian micro-fossil represents the size of a nucleus, this suggests a living Martian organism with a diameter of around 400 nm. Scaling for the largest example of a mitochondrion would suggest a Martian organism with a diameter at least twice as great – about 800 nm.

The lysis (i.e. the disruption and death) of terrestrial bacterial cells releases the cell contents into the adjacent environment. If Martian micro-organisms were cells with non-rigid walls, lysis would cause a cell's contents to spill out into the surroundings, in the way that a small droplet of ink dropped on to blotting paper makes an ink-blot. The ratio of the original cell size to that of the aureole (the equivalent of the blot) will depend upon the viscosity of the cell fluid and the absorbency of the surrounding medium. If these factors can be reliably estimated, in principle it should be possible to get some idea of the size of the original cell. For normal fluids, a droplet placed on a highly absorbent surface may give rise to a final aureole up to ten times larger than the droplet itself. A Martian cell of 400–800 nm might on lysis therefore produce a chemically 'stained' aureole 4–8 μm across. If the possible fossil organisms in ALH 84001 are indeed remnant nuclei or large organelle-like structures, a trace of the distribution of PAHs across each fossil might be expected to show PAH concentration rising to a peak as the fossil is approached and then falling again. Resolution on the 50–100 nm scale may make it possible to determine their nature with certainty.

An alternative possibility is that some form of process involving reduction in size during periods of long-term starvation may have operated to reduce the dimensions of the Martian micro-organisms to a few per cent of their normal value. If these organisms died while in their shrunken state, their volume or mass could be expected to decrease still further through total dehydration to something closely similar to those observed. Thus we could be looking at fossilized 'microbial cadavers' rather than fossils of complete viable micro-organisms.

So – Have We Found Life on Mars?

No single piece of the evidence presented by McKay and his colleagues is by itself more than a persuasive indicator that traces of ancient biological activity exist within ALH 84001. Even the full assemblage of data does not constitute cast-iron evidence for the possible presence of life on Mars, but what is undeniable is that the relationship and pattern of association of possible biological indicators within ALH 84001 speak uniquely of the possibility that life was once present on Mars. At present the case for life on Mars is finely balanced, with a bias slightly towards the NASA interpretation. The problems that remain may be resolved only by the collection of a substantial quantity of samples from the surface, and subsurface, of Mars over a wide range of geological terrains. Ultimately, it may be necessary for astronauts to go there to carry out a comprehensive search for evidence of life. Only then will we know for certain that life has existed, and may still thrive on Mars, and possibly elsewhere beyond Earth. As William Schopf of the University of California, a renowned authority on ancient terrestrial fossil micro-organisms, has said, 'exceptional claims require exceptional evidence'. For the moment the existence of life beyond the Earth remains an exciting possibility, but looks increasingly probable.

Notes

1. The classification SNC is derived from the initial letters of three meteorite types. S stands for shergottite, the name given to the type of meteorite that fell in the Indian state of Bihar near the town of Shergotty in 1865; N is derived from the type name nakhlite, given to the meteorite that fell at Nakhla, Egypt, in 1911; the C is for chassignite, named for the meteorite that fell in Chassigny, France, in 1815.
2. 'Search for past life on Mars: Possible relic biogenic activity in Martian meteorite ALH 84001,' by D. S. McKay *et al.* This is the key scientific paper that appeared in the August 16, 1996, issue of *Science*, Volume 273, pp. 924–930. It is inevitably rather technical, but is essential reading.

Galileo – A Year Among Jupiter's Satellites

DAVID A. ROTHERY

In January 1610 the Italian astronomer Galileo Galilei trained his telescope on Jupiter and saw that it was accompanied by four tiny points of light. Within a few days, Galileo had determined that these points continually changed their relative positions and yet remained close to Jupiter in the sky. He correctly deduced that they are in orbit around Jupiter, and realized that this threw a fatal spanner in the works of the then current theory of the Solar System, which held that all motion was centred on the Earth. As is well known, Galileo's subsequent championing of the heliocentric (Sun-centred) theory of the Solar System brought him into head-on conflict with the religious authorities.

Galileo had discovered the four largest of Jupiter's satellites, which are today known as the Galilean satellites. Some basic information on them is given in Table 1. The smallest, Europa, is slightly smaller than our own Moon, whereas the largest, Ganymede, is bigger than the planet Mercury. Being of fifth or sixth magnitude when Jupiter is at opposition, they can be seen in small telescopes or even with steadily-held binoculars. We now know that Jupiter has twelve other satellites, but they are all very much smaller than the Galileans. The first of these to be found, Amalthea

Table 1. Basic data for Jupiter's Galilean satellites.

Name	*Radius (km)*	*Mass (10^{20} kg)*	*Density (10^3 kg/m^3)*	*Orbital period (days)*	*Orbital radius (10^3 km)*
Io	1815	894	3.57	1.77	422
Europa	1569	480	2.97	3.55	671
Ganymede	2631	1482	1.94	7.16	1070
Callisto	2400	1077	1.86	16.69	1883

(a thirteenth-magnitude object only 180 km across), was discovered as recently as 1892 by Edward E. Barnard using the 36-inch (91-cm) Lick refractor in California.

The Galilean satellites constitute what has been described as a 'solar system in miniature', and display a decrease in density from the innermost to the outermost that is reminiscent of the decrease in density of the planets working outwards from the Sun. The explanation is probably the same: a temperature gradient. It is believed that heat emitted from Jupiter while it and its satellite system were forming (during the birth of the Solar System four and a half billion years ago) allowed only denser, less volatile material to condense close to the planet. The outer two Galilean satellites, Callisto and Ganymede, are the largest, and their icy surfaces demonstrate cold conditions not only today (Jupiter's satellites all have surface temperatures of about −170°C) but also during their birth. Furthermore, their low densities show that they consist of about 50% ice mixed with rock or other denser material. The innermost Galilean, Io, is a rocky body a little larger and denser than our own Moon. It is totally ice-free, indicating that the temperature was too high for ice to condense this close to Jupiter when Io was being formed. Europa belongs to an intermediate category, having an ice-covered surface but a density sufficiently high to demonstrate that the ice can be no more than about a hundred kilometres thick.

That much was known before the first space missions to the Jupiter system. However, when NASA's probes Voyager 1 and Voyager 2 flew through the Jupiter system in 1979, the images transmitted back to Earth showed the Galilean satellites to be worlds dramatically different to each other in appearance, and far more fascinating than the bald facts would suggest. They were expected to resemble our own Moon, which geologically has been virtually dead for about three billion years, since the last eruptions of basalt lava into the mare basins (the lunar 'seas') occurred. Since then, nothing much has happened on the Moon other than the gradual accumulation of impact craters, formed when lumps of cosmic debris hit the surface at speeds of a few tens of kilometres per second. It was widely expected that the Galilean satellites would each have similarly ancient, heavily cratered surfaces. This turned out to be true only for Callisto. No impact craters at all were identified on Io, whose surface was dominated by volcanic landforms, including several volcanoes that were actually erupting during the Voyager fly-bys. Europa's surface was found to have very few impact craters, but to show abundant signs of recent deformation. Ganymede was fairly heavily cratered, but the craters were too few to hide the signs that its surface has been sliced across by processes of eruption and deformation.

Thus, the Galilean satellites display a trend in geological activity, the outermost being the least active and the innermost the most active. For the surfaces of Io, Europa and (possibly) Ganymede to be younger than the Moon's, they must have a supply of internal heat that the Moon lacks. Virtually all the Moon's heat comes from the decay of long-lived radioactive isotopes of potassium, thorium and uranium. The same is true of the Earth, but the Earth's volume is so much greater relative to its surface area that it remains hot inside, whereas the Moon has lost most of its heat. Io is not very much bigger than the Moon, and for its active volcanoes to be powered by radioactive heat alone the satellite would have to contain a ludicrously high concentration of radioactive elements.

In fact, Io's main power source is tidal energy. Jupiter is by far the most massive of the planets, and has a very strong gravitational field. The varying gravitational force experienced by Io as it orbits Jupiter distorts its shape and generates heat by a kind of internal friction. This is what powers the volcanoes. Io experiences such dramatic tidal heating only because it belongs to a family of satellites. Io's orbital period is exactly half that of Europa, so Io overtakes Europa at exactly the same point in each orbit. The slight but repeated gravitational tug between the two satellites keeps their orbits elliptical. Without it, tidal forces from Jupiter would drag Io into a perfectly circular orbit in a few million years, and tidal heating would then cease. The varying tidal stresses in Europa provide it too with a source of tidal heat, and tidal heating in the distant past is probably the cause of Ganymede's complex surface history.

It is fitting that the first space probe to be sent into orbit around Jupiter should bear Galileo's name. Galileo was launched from the payload bay of a space shuttle in October 1989. Because its solid-fuel rocket engine was not powerful enough to send it directly towards Jupiter, Galileo was sent past Venus, back past the Earth, out into the asteroid belt (where it obtained the first close-up images of an asteroid, Gaspra), and back in past the Earth again. Only then, after three close passes near planets, had the probe picked up enough speed (through the gravitational sling-shot effect) for it to be sent out to Jupiter. Passing through the asteroid belt, Galileo imaged the asteroid Ida and discovered its tiny satellite Dactyl, and eventually reached Jupiter in December 1995. Shortly before arrival, a small probe was sent ahead to enter Jupiter's atmosphere and obtain the first direct analyses of its composition. The main craft was swung close to Jupiter so that it entered an elongated orbit around the planet. By a combination of thruster bursts and sling-shot passes close to the satellites, Galileo's initial two year orbital tour was designed to visit each of the Galilean satellites in turn. The complicated orbit, sometimes referred to as a petal plot, is shown in Figure 1, and the details of close encounters with the satellites are given in Table 2.

The Galileo mission will undoubtedly go down as a triumph in the history of space exploration, but it has not been without its hitches. The first came when Galileo's main high-gain parabolic antenna for communication with the Earth failed to unfold, despite repeated

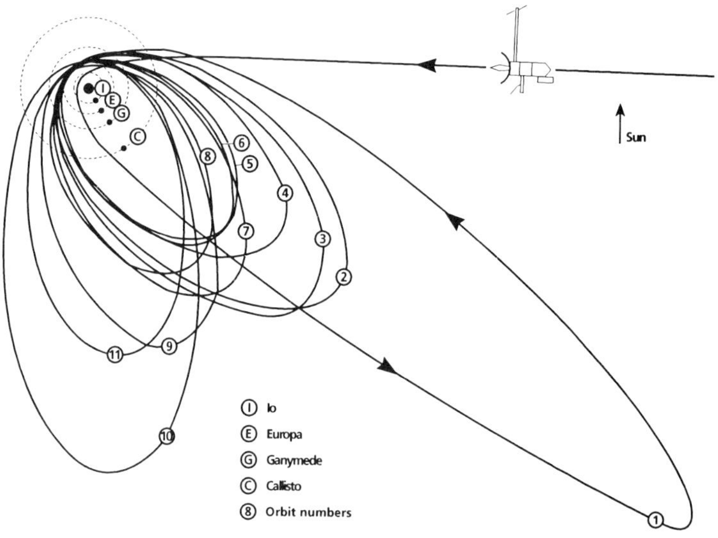

Figure 1. Galileo's 1996–7 orbital tour of the Jupiter system. After orbit 11 the Galileo Europa Mission will begin, with a series of eight close passes by Europa during 1998, followed by a dive inwards for a close look at Io in 1999. This will be a 'suicide mission' because the high levels of radiation close to Jupiter would destroy Galileo's electronics, but hopefully not before some very detailed images had been sent back.

Table 2. Encounters between Galileo and the Galilean satellites during the primary mission.

Date	*Satellite*	*Closest approach (km)*
1995 December	Io (during orbital insertion)	1400
1996 June	Ganymede	835
1996 September	Ganymede	262
1996 November	Callisto	1118
1996 December	Europa	692
1997 February	Europa	587
1997 April	Ganymede	3059
1997 May	Ganymede	1585
1997 June	Callisto	416
1997 September	Callisto	524
1997 November	Europa	1125

attempts to unjam it. As a result, the images and other data collected had to be sent back using a low-gain antenna, which had a much lower capacity. In order to make best use of this, new data compression routines were transmitted to Galileo's on-board computer during the flight to Jupiter. Even so, data had to be transmitted to Earth much more slowly than originally anticipated, to allow the incredibly weak signal to be detectable above the level of noise. A further disappointment came shortly before arrival at Jupiter, when the on-board tape recorder showed signs of unreliability, and so it was decided to run it at low speed only. Top priority was given to recording and relaying data from the Jupiter atmospheric entry probe, so unfortunately no images could be recorded of the only scheduled close encounter with Io. Volcanologists – myself included – were disappointed, but were pleased with the quality of the more distant images of Io that were obtained on each subsequent orbit. What follows is a summary of the results from Galileo's first year at Jupiter.

Io

Although Galileo's imaging system was turned off during its only close encounter with Io, other instruments were running. Magnetometer measurements showed that Io generates a strong magnetic field, which had not been detected for any satellite before. This field is probably generated in a core of molten iron (similar to the Earth's), and the fact that Io has a dense core was independently established by using Io's gravitational pull on Galileo to determine how the satellite's density varies with depth.

If there were any lingering doubts that Io is a volcanically active world, then the first Galileo images dispelled them. New eruption plumes were seen, where sulphur and sulphur dioxide are jetted into space. Bright, fresh deposits from such plumes could be seen on the surface that were not there 17 years previously during the Voyager encounters; some eruption-plume deposits that had been prominent in Voyager images had by now faded into the background. Io's recent volcanic activity is not confined to explosive eruptions, because new lava flows can be seen too.

Figure 2 shows a crescent view of Io as seen from Galileo, revealing an eruption plume from the volcano Ra Patera. The comparison between Voyager and Galileo images of the Ra Patera area included in the same figure shows notable changes in the distribution of eruption-plume deposits (pale) and lava flows (dark). The major brightening around Ra Patera in the Galileo image had previously been noted on images from the Hubble Space Telescope obtained in July 1995, but was not present in March 1994, which brackets the main eruption into a 16 month period.

Figure 3 compares Voyager and Galileo images of the region around the volcano Prometheus. This volcano (at the top of each view) is surrounded by a ring of pale eruption-plume deposit, which has changed shape in the 17 years between the two images. There also appears to be a new dark lava flow extending eastwards from the volcano's summit, and there are changes near the volcano Culann Patera, at the lower left.

Ever since the Voyager fly-bys, geologists have been debating whether the lava flows on Io are composed of sulphur (as Io's generally yellow-red colour suggests) or whether they are produced by molten rock (such as basalt, which is common on the Earth and Moon). Galileo carries the most sophisticated spectrometer ever deployed in the outer Solar System, which is being used to analyse how the reflectance of sunlight from areas of Jupiter's satellites

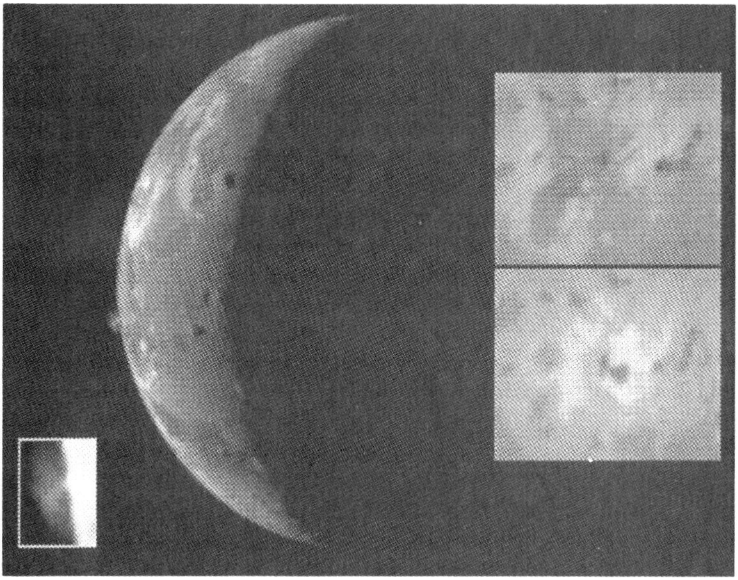

Figure 2. A crescent view of Io, as seen by Galileo in August 1996. An eruption plume on the limb emanates from the volcano Ra Patera. This plume is seen enlarged and enhanced in the small inset at lower left. The inset at the right compares a Voyager 1 (1979) view (top) with a Galileo view (bottom) of identical areas, centred on Ra Patera (NASA P-47209.)

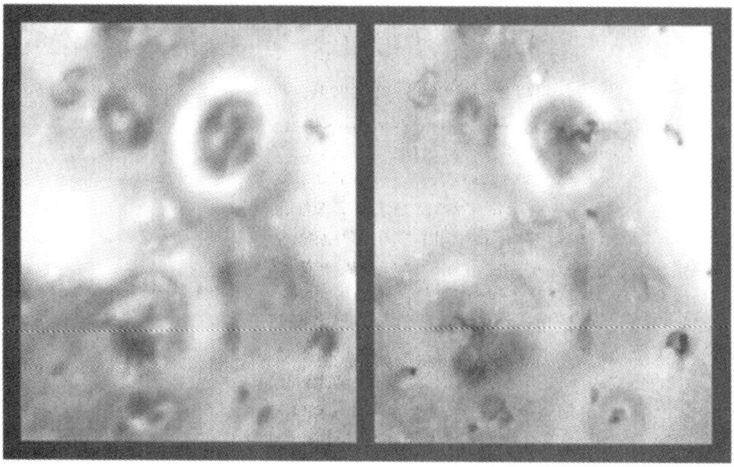

Figure 3. A comparison between Voyager 1 (December 1979, left) and Galileo (September 1996, right) images of the region around the volcano Prometheus. (NASA P-47972.)

varies with wavelength. We already know that there is a lot of sulphur involved in the explosive eruptions, and we would expect sulphurous coatings to accumulate quickly, even on basalt lava flows. So, although spectrometer studies of reflected light are a good way of determining the surface composition, they cannot distinguish a thin (micrometre-thick) coating of sulphur on the surface of a basalt flow from a flow that is entirely composed of sulphur.

The best hope of settling the sulphur versus rock controversy on Io will come from analysis of the infrared radiation emitted by flows while they are actually being erupted. Molten rock is considerably hotter than molten sulphur, although the situation is complicated because we know from studies of active terrestrial lava flows that most of their surface is usually covered by a chilled crust, and the proportion of molten material visible from above is usually very small. Infrared monitoring of Io's eruptions has been carried out for over a decade using Earth-based telescopes, and analysis of these observations and the preliminary infrared data from Galileo suggest that at least some of Io's active lava flows are too hot to be sulphur, and are therefore likely to be true molten rock.

Europa

Europa was the most poorly imaged of the Galilean satellites by the Voyager probes, so there was great excitement among planetary scientists at the prospect of better coverage by Galileo. Each of Galileo's three scheduled close encounters with Europa was during the second year of its orbital tour, and so came too late for this report. However, more distant Galileo images of Europa obtained during the first year provide tantalizing glimpses, and confirm the Voyager view of Europa as a world of subdued relief with a young icy surface cracked and refrozen in a complex series of events (Figure 4). Europa's history of cracking is most clearly demonstrated by the narrow bands (mostly less than about 20 km wide), some of which run for thousands of kilometres across its surface. Vast regions of Europa look like pack ice in the Arctic Ocean, with bright ice floes separated by dark refrozen cracks. It can be fascinating to do the 'Europa jigsaw puzzle' by mentally fitting the bright areas back together in the configuration they must have had before they drifted apart.

The most detailed view obtained during Galileo's first year (Figure 5) shows that the youngest dark bands are slighly elevated

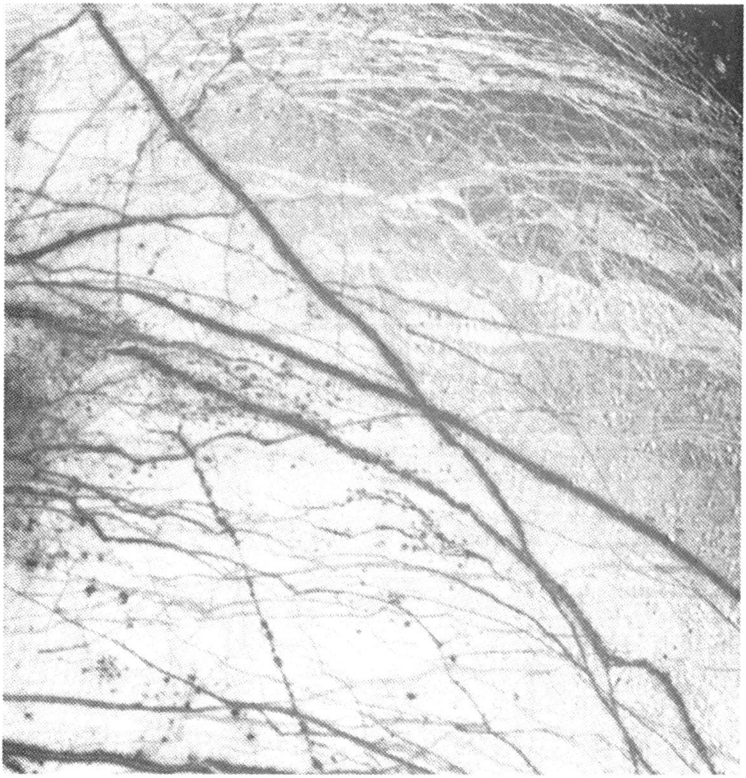

Figure 4. Galileo image of part of Europa's northern hemisphere 1260 km across. The original is in false colour (made by combining images recorded at 559, 757 and 989 nm), which reveals subtle variations within the dark bands and considerable colour variation in the paler terrain in between.

features, and that they may subside with age. The bright terrain between the dark bands is not totally smooth, but shows hints of even older ridges. Evidently, Europa's recent history has been very complicated.

Any resemblance to the Earth's Arctic Ocean is superficial, because Europa's surface temperature is about −170°C, and ice this cold is extremely rigid and must behave very differently to the much warmer polar ice on Earth. On the other hand, tidal heating is likely to keep Europa's interior warm, and ever since Voyager it has been

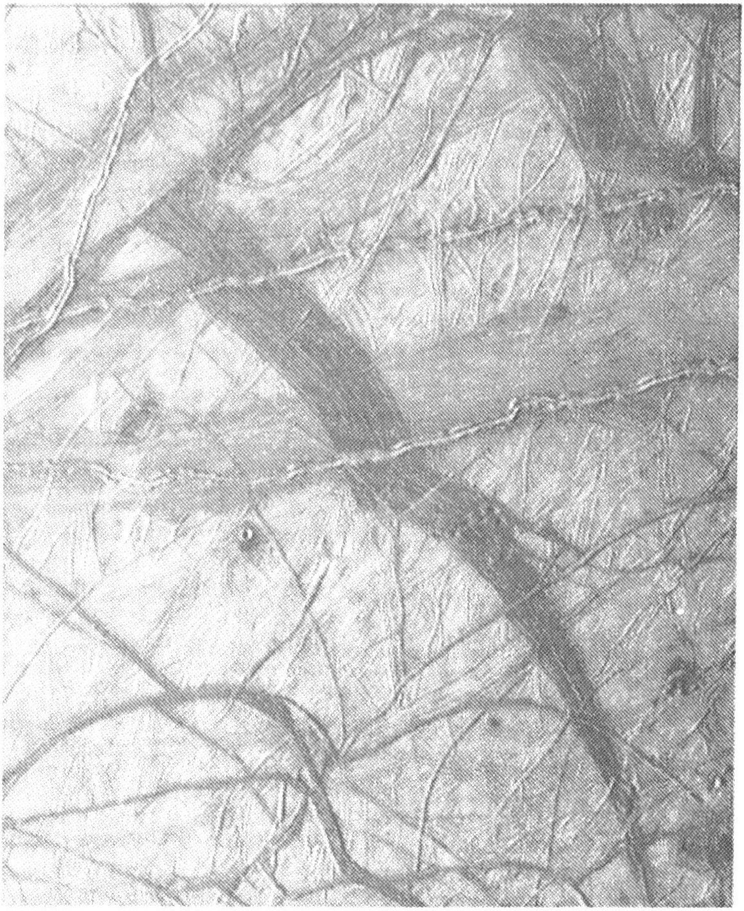

Figure 5. Galileo image of part of Europa, showing an area 160 km across. This detailed view, obtained with the Sun low in the sky (to the left), exaggerates Europa's subdued topography, and reveals an amazingly complex geological history, even in the bright terrain, which was previously thought to be smooth. The surface is dominated by a series of low ridges, probably marking sites where fresh ice welled upwards during fracturing events. The double ridge near the top left marks one of the youngest of these. The dark wedge-shaped band running diagonally from top left to bottom right is one of the oldest clearly visible features, as can be deduced from the fact that it is cut across by several narrower ridges. Only one (2 km diameter) impact crater is visible (below left of centre), demonstrating the young age of the surface. (NASA P-48127.)

speculated that heat escaping from Europa's rocky core could be sufficient to melt the base of the overlying ice. The motion of the platelets of surface ice indicated by the patterns of dark bands shows that melting of the lower ice has almost certainly happened fairly recently, at least on a local scale and for limited periods of time, and the possibility remains that, sandwiched between a few tens of kilometres of ice and its rocky interior, Europa has a persistent global water ocean. If so, water is probably circulating through the outermost layers of rock, and emerging back into the ocean at hot springs charged with chemicals dissolved out of the rock.

This is a particularly exciting prospect, because we know that life can survive around hot springs on the Earth's ocean floors, where primitive bacteria feed on chemical energy. These 'chemosynthetic bacteria' provide the basis of a food chain that is separate from the main food chain of life on Earth, which is based on photosynthesis – the process in which green plants derive energy from sunlight. If chemosynthetic bacteria were transported from Earth to a new home beside a hot spring under Europa's ice, they would be snug enough there to survive and even flourish. Intriguingly, recently genetic evidence has begun to show that life on Earth probably began at hot springs rather than in the 'primaeval soup' of popular legend. If on Earth (and apparently on Mars), why not then on Europa? Hot-spring communities on our own ocean floors include animals as advanced as clams, shrimps and crabs that feed off the bacteria; and although it seems unlikely that Europa harbours such complex organisms, the odds seem very much in favour of simple life forms there.

Ganymede

As befits the Solar System's largest planetary satellite, Ganymede had plenty of surprises in store. Its density is much lower than Europa's, showing that its rocky interior must be buried beneath several hundred kilometres of ice. Galileo's first big surprise was the discovery that Ganymede has its own magnetic field. This was more surprising than for Io, because Ganymede's surface is everywhere pockmarked by impact craters, so we know that there can be little, if any, widespread internal activity today, except very deep down. However, it now seems that Ganymede's interior must contain an electrically conducting fluid, whose motion generates the magnetic field. Gravity models of Ganymede's interior (derived by tracking Galileo) suggest that this is probably in the form of a molten iron core like the Earth, Io and (probably) Europa, although a layer of salty water or warm convecting ice at depth are alternative possibilities.

The wonderful detail revealed by Galileo images of Ganymede's

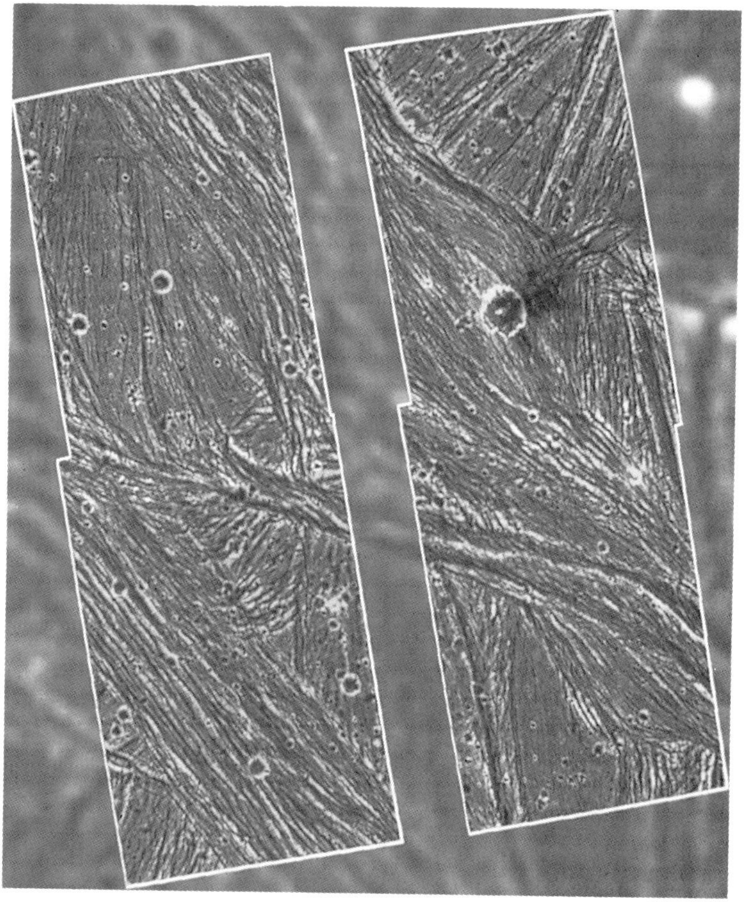

Figure 6. A mosaic of four Galileo images of the Uruk Sulcus region of Ganymede, superimposed on a Voyager image of the same region. The Galileo images have a pixel size of 74 metres across, and show much finer details than the Voyager image, which was recorded with 1.3-kilometre pixels. The area shown is 110 km across. Note that younger belts of ridge-and-groove terrain cut across older belts, but that superimposed impact craters demonstrate that the tectonic activity responsible for the ridges and grooves has long ceased, probably more than 1 billion years ago. (NASA P-47064.)

surface is exemplified by Figure 6. Voyager images had shown that large tracts of Ganymede's surface are composed of sets of parallel ridges and grooves. High-resolution Galileo images look much the same, and if you don't know the scale it is virtually impossible to tell such an image from a low-resolution Voyager image. Whatever process has carved up Ganymede's surface, it has done so at a variety of scales, and it is likely that the ridge-and-groove terrain is a result of fracturing and deformation. The close spacing revealed by Galileo between fine-scale ridges and grooves indicates that the surface layer that has been deformed is probably thinner than had previously been expected, and is possibly only about a kilometre thick. There must have been a more ductile, less brittle, layer below, but there are no indications of this softer material having reached the surface. For example, no signs of icy volcanism (such as lobate flow fronts or embayment of topography by floods) were apparent in Galileo's first year's crop of Ganymede images.

Galileo's orbital profile path enabled it to image some parts of Ganymede from different vantage points in successive encounters. The stereoscopic view obtained by using two such images enabled the topography to be mapped using computer image-processing programs. This three-dimensional information can then be used to construct perspective views, such as the one shown in Figure 7.

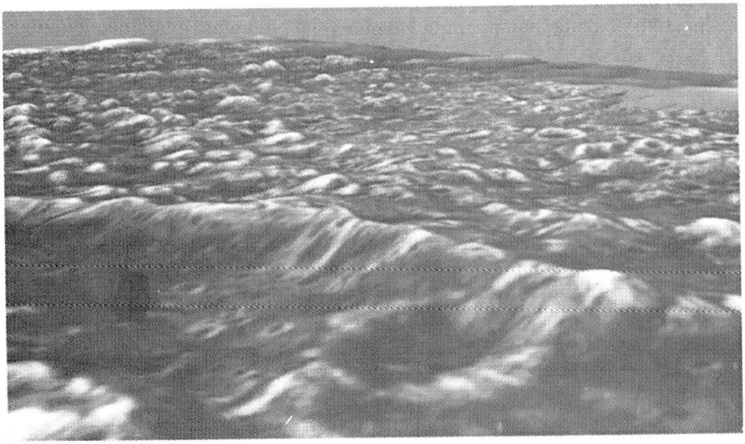

Figure 7. Perspective view of part of Ganymede, made by combining Galileo images obtained from different vantage points.

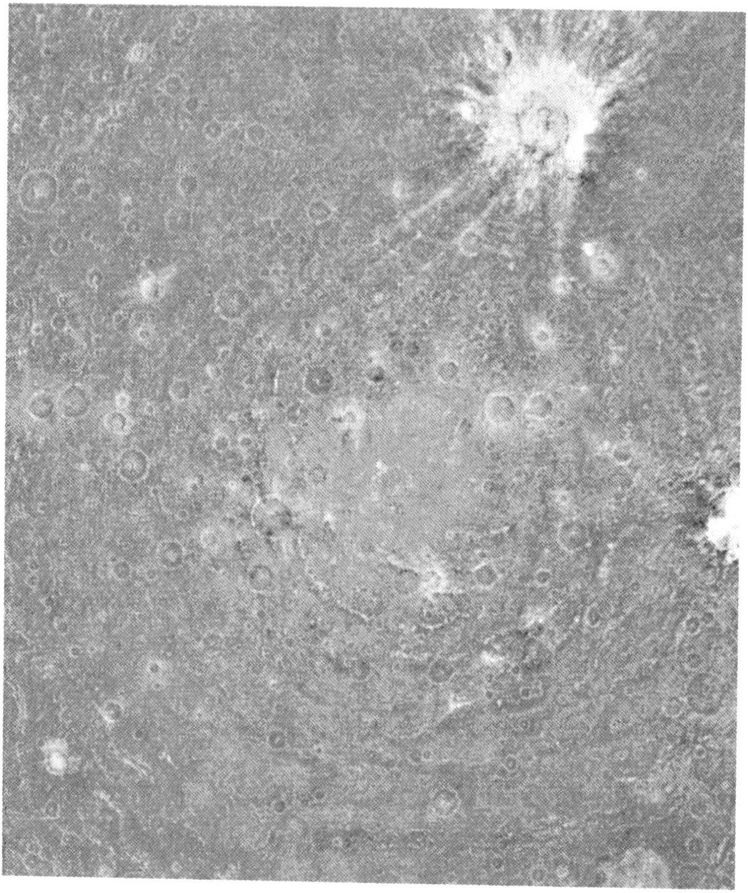

Figure 8. A 1200-km wide mosaic of Galileo images showing the multi-ringed impact basin Asgard on Callisto. The site of a very large ancient impact is indicated by concentric rings of fractures. Several younger impact craters are superimposed upon these fractures, the youngest being apparent because of its bright ejecta blanket and rays. (NASA P-48126.)

Callisto

Voyager showed Callisto to be very heavily cratered, and therefore to have ceased geological activity long ago. Craters are crowded so closely together that no traces have survived of what the surface was like in Callisto's younger, warmer and more active youth. However, the cratering record is in itself fascinating because the frequency of crater sizes does not match that seen on the Moon, demonstrating that the Jupiter system was bombarded by a population of impactors different to that which afflicted the inner Solar System during the late-heavy bombardment (soon after the formation of the Solar System), and which is recorded by the cratering of the lunar highlands. On the other hand, Callisto has multi-ringed basins in excess of a thousand kilometres across, testaments to impacts by asteroid-sized bodies, similar to those on the Moon (like the Imbrium Basin) except that they were not subsequently flooded by mare-forming lavas (Figure 8).

Another intruiging aspect of Callisto's cratering is that there are several chains of craters (Figure 9). Many of them may be the result of collisions with tidally broken comet nuclei, like that of Comet Shoemaker–Levy 9 which struck Jupiter in July 1994. Despite the abundance of large craters, the highest-resolution images show that Callisto is distinctly lacking in craters less than a few hundred metres across. It would be unreasonable to suppose that Callisto has not been bombarded by impactors sufficient to make such craters, so the logical conclusion is that craters this small are being continually removed. One possible explanation is that solar ultraviolet light and cosmic rays break the ice into hydrogen and oxygen gas, which escape to space, thereby concentrating the rocky or carbonaceous impurities as a dark, dusty layer.

In contrast to the three layer ice-rock-iron structure revealed for Ganymede, Galileo's gravity mapping of Callisto showed no signs of internal layering. Callisto therefore appears to be an undifferentiated, homogeneous mixture that has never been heated enough to allow it to settle out into layers.

The future

Planetary scientists eagerly await the results of Galileo's second year at Jupiter, and then the extended Galileo Europa Mission in 1998. As more and more images arrive, we shall be able to interpret the first scattered detailed views (described above) in their proper global contexts. It will be particularly fascinating to discover whether the fine-scale ridges and grooves discovered on Europa's bright terrain (Figure 5) are widespread, and to investigate whether they were formed by the same processes as those responsible for Ganymede's much older ridge-and-groove terrain (Figure 6).

Figure 9. This amazingly detailed Galileo image of Callisto shows an area only 13 km across. The irregular trough running obliquely across the picture is part of a chain of overlapping craters. Pale, fresh ice has been exposed on the walls, apparently as a result of downslope mass-wasting processes. The subdued topography in the foreground appears to be similarly mantled by a dark deposit. (NASA P-48124.)

To find out more

Internet users can find updates on the Galileo tour of the Jupiter system and a large number of images at:

http://www.jpl.nasa.gov:80/galileo/

The satellites of Jupiter and of the other giant planets are described in detail in *Satellites of the Outer Planets: Worlds in Their Own Right*, by David A. Rothery, Oxford University Press (1992).

Cassini/Huygens – Mission to Saturn and its moon Titan

PAUL MURDIN

I watched the launch of the European Space Agency's Ariane flight 501 from Kourou in June 1996 and saw it explode, depositing into the Guyanese swamp the flaming remains of four satellites (the Cluster mission) equipped with a considerable amount of British space instrumentation. I woke up one Sunday morning in December 1996 to hear that Mars 96, a Russian space mission containing three British pieces of instrumentation, had gone off course and was expected to drop into the Pacific Ocean later that day. So I know that space launches are hazardous and there is a risk attached to making predictions about them. On the other hand, late in 1995, ESA's Solar and Heliospheric Observatory (SOHO) was launched and, several weeks early due to the accuracy of its launch, it arrived on station, from where it has been observing the Sun continuously and with great success. And ESA's Infrared Space Observatory (ISO) has been orbiting the Earth, making infrared observations with great effectiveness since its launch at the end of 1995.

By and large, launchers fail their space missions about 1–3% of the time. This is not a good enough statistic to tempt me to take a joyride on a rocket, but it is good enough to gamble on predicting, as I write this article early in 1997, that by the time you read it in the *1998 Yearbook of Astronomy*, NASA will have successfully launched the Cassini space mission towards Saturn and its system of rings and satellites. In 2004, this will result in the first detailed look at Saturn since the Pioneer spacecraft of 1979 and the two Voyagers of 1980 and 1981.

The mission includes the ESA space probe Huygens, piggybacking on Cassini to Saturn. Huygens is destined to drop on Titan, Saturn's main moon, and explore it for a few minutes. Both Cassini and Huygens carry British instruments, so British space scientists have been scrutinizing the planning of the mission for fifteen years. As rational people they therefore trust the engineering. Even so, with the rocket being readied for launch, and for the extra effect that their actions might have, some of them might have crossed their fingers or rubbed lucky charms!

If everything at launch – and the subsequent seven years! – goes to plan, the Huygens mission will be the first time that Europe has landed a spacecraft on another celestial body, and only the second time (after the NASA Galileo probe to Jupiter in 1996, which contained British equipment to examine Jupiter's atmosphere) that British instruments have tested the conditions on another planet by going there.

The mission objectives

When Voyager 1 sent back images of Titan, Saturn's 5150-km satellite (bigger than the planet Mercury), astronomers could see very little. Instead of craters, ice sheets or volcanoes, the images showed a hazy orange fog which shrouded the surface. The surface has never been seen. Titan's atmosphere is nitrogen, rendered opaque by the presence of methane and carbon compounds. Methane is changed to ethane under the action of sunlight, and the methane in Titan's atmosphere must come from somewhere – a methane ocean, perhaps, or underground springs. In many ways Titan's atmosphere resembles how scientists imagine the prebiotic atmosphere of the Earth to have been, which makes it especially interesting to us as human beings. Titan gives us a glimpse of how Earth might have been before life started here, and generated the oxygen-bearing atmosphere which we animals breathe.

In 1982 a mission to Titan was proposed to ESA by Paris astronomer Daniel Gautier, who was one of the Voyager scientists. It was accepted and designated M1, the first Medium-sized mission of ESA's Horizon 2000 programme, in conjunction with a mission to Saturn which NASA was planning to follow up the Voyager discoveries in more detail. The joint mission is designed to make a detailed study of Saturn, its rings, its magnetosphere, its icy satellites – and Titan.

The launch

The nominal mission profile is given in Table 1. The scheduled launch window, from October 6, 1997 to November 4, 1997, provided a 30-day launch period (nominal was 09:40 GMT, October 6). If Cassini/Huygens was launched between these dates, with the Earth in the correct arc of its annual orbit, there will have been enough fuel in the rocket to carry out all the manoeuvres required in

Table 1. Mission profile (nominal).

Mission event	*Start date*	*Days from launch*
Launch on Titan IV launch vehicle	Oct. 6, 1997	0
Aphelion 1 (furthest distance from Sun in first orbit)	Nov. 1, 1997	26
Perihelion 2 (closest approach to Sun)	Mar. 23, 1998	169
Venus 1 fly-by	Apr. 21, 1998	198
Deep space manoeuvre to target Venus 2	Dec. 2, 1998	423
Aphelion 2 (furthest distance from Sun in next orbit)	Dec. 4, 1998	424
Venus 2 fly-by	June 20, 1999	622
Perihelion 2 (closest distance to Sun)	June 27, 1999	629
Earth fly-by	Aug. 16, 1999	680
High-gain antenna can be used from now on	Jan. 29, 2000	696
Jupiter fly-by	Dec. 30, 2000	1181
Science observations begin at Saturn	Jan. 1, 2004	2277
Phoebe fly-by	June 12, 2004	2441
Saturn orbit insertion manoeuvre	July 1, 2004	2460
Manoeuvre to target Huygens probe to Titan	Sep. 12, 2004	2533
Huygens separates from Cassini to go to Titan	Nov. 6, 2004	2588
Manoeuvre by Cassini to target Titan fly-by	Nov. 8, 2004	2590
Cassini at Titan (4 hours)	Nov. 27, 2004	2609
Huygens parachutes onto Titan (120–150 min)	Nov. 27, 2004	2609
Operations on surface of Titan (3–30 min)	Nov. 27, 2004	2609
End of nominal mission (after 4-year tour)	July 1, 2008	3921

the first hours and set the space probe accurately on its seven-year flight. If for some reason (e.g. poor weather at Cape Canaveral) it could not be launched then, but was still launched before November 15, 1997, Cassini will have had to use more of its propellant just to get to Saturn, and will not have enough to carry out all the manoeuvres planned for it at Saturn itself – some science will be lost.

If for some reason NASA's rocket did not do its stuff in October or November 1997 but did not blow up, all is not lost. There is a not-quite-so-good second chance at the end of 1997, and a not-quite-as-good-as-that third chance in March 1999; the arrival times at Saturn are several years later (Table 2).

The launch rocket for Cassini/Huygens was the Titan IV/Centaur system, the largest and most powerful American expendable launch vehicle. The Titan IV, together with two solid-motor strap-ons and a Centaur upper stage, weighed a total of 940,000 kg, of which 840,000 kg was propellant. Cassini/Huygens has a mass of 5650 kg at launch, half of it on-board fuel, and 6% of it the Huygens probe. Think of Huygens as a small car like a Fiat Panda piggybacking on a small bus (Cassini). The mass of Cassini/Huygens (to be compared with about 825 kg for Voyager) makes it one of the heaviest spacecraft ever to be launched on an interplanetary mission.

Some people believe that this kind of massive space vehicle is becoming obsolete, as NASA and ESA turn to spacecraft which are 'smaller, faster, cheaper' (the phrase is Dan Goldin's, the Chief Administrator of NASA). There are several reasons why space agencies are moving towards smaller spacecraft. Financial pressure is one – the more bits and pieces there are in a spacecraft, the more massive and expensive it is. Time pressure is another – it is unusual for space missions to take less than a decade, and they can last for 25 years. Few people want to work on a project that lasts this long, especially not a student facing a three-year Ph.D. There is also a growing realization that technology now envisages more intelligent and therefore more autonomous spacecraft. (Cassini goes some way towards autonomy – it has to. Radio waves take about 80 minutes to travel between Saturn and Earth. There would be a three-hour interval between a call for help from Cassini, and a reply from Earth telling it what to do.) So Cassini might be among the final missions of this size, 'the last of the dinosaurs'.

If all went well in the October launch, the first three stages and part of the fourth stage (the Centaur) burnt for about 12 minutes to put the Centaur/Cassini stack in an Earth parking orbit. The stack

Table 2. Launch windows.

	Gravity assists in trajectory	*Launch period*	*Arrival date*
Primary	Venus–Venus–Earth–Jupiter	Oct.–Nov. 1997	July 1, 2004
Secondary	Venus–Earth–Earth	Dec. 1997–Jan. 1998	Oct. 13, 2006
Back-up	Venus–Earth–Earth	Mar.–Apr. 1999	Dec. 22, 2008

then coasted for about 15 minutes until the Centaur stage was reignited for 8 minutes. The Centaur stage prepared to separate from Cassini over the next 10 minutes, while the spacecraft was activated to operate on its own. Shortly after separation, the Centaur performed a small manoeuvre to deflect itself from the spacecraft's path to avoid the possibility of a collision between the inert Centaur and the spacecraft.

Gravity assists

Even the Titan IV/Centaur rocket was not powerful enough to send Cassini/Huygens on a direct path to Saturn, so far away. The rocket gave the spacecraft a speed of 4 km/s, and it would take a speed of 10 km/s to get directly to Saturn. But 4 km/s can get the spacecraft to Venus, and this is where Cassini/Huygens will be heading at the beginning of 1998, to use the technique of 'gravity assists' to pick up speed. In April 1998, Cassini/Huygens will fly past Venus on an overtaking orbit. The gravitational pull of Venus will tug the spacecraft along, accelerating it and giving it more speed. Of course, someone has to pay for this extra energy, and it is Venus. Just as Venus pulls the spacecraft, the spacecraft pulls Venus and Venus will lose some of its 35 km/s orbital speed. But since Venus is so much more massive than the spacecraft, the spacecraft gains a lot of energy, whereas Venus won't notice what it has lost.

Even this first gravity assist is not enough: Cassini/Huygens will cross the orbit of Venus and catch the planet a second time in June 1999, stealing a little more speed. From there, it zips back to the Earth and gets a third gravity assist in August 1999. (The encounter will reduce the speed of the Earth in its orbit and delay the start of the millennium celebrations by about 10^{-20} seconds.) This third gravity assist sets Cassini/Huygens on its way to the outer Solar System, and it flies by Jupiter in December 2000 for its last 'fix' of speed. After this it is mostly plain sailing to Saturn.

This is the most complex gravity-assist trajectory yet attempted for a deep-space probe. It is hard to get everything right by dead-reckoning, so the flight plan allows for trajectory correction manoeuvres (TCMs) to maintain the spacecraft on its planned trajectory, nudging it with small adjustments, without which it would miss Saturn by many millions of kilometres. The TCMs around each gravity-assist manoeuvre typically require three nudges: the first after the encounter, to compensate for any error, and the second and third to deliver the spacecraft as accurately as possible to the next encounter. In addition to these manoeuvres there is a large deep-space manoeuvre between the two Venus encounters.

Hot and cold

Because Cassini/Huygens will take a roundabout route to Saturn, it is exposed to large changes in solar radiation. During early orbits it will be subject to more than twice the intensity as at the Earth, and in the vicinity of Saturn only about one-hundredth. This places varying demands on the power and thermal management of the spacecraft. Radioactive generators will supply electrical power for Cassini and warm the equipment slightly in the deep cold of the environment at Saturn. Huygens relies on Cassini for the seven-year voyage to Saturn because it has batteries which will last only for the last 153 minutes of its drop to Titan. Huygens has reflective thermal insulation to stop it overheating when it is near the Sun. A large radio dish, the 4-m diameter high-gain antenna made in Italy, will start its active life in a low-tech way, as a sunshade for the combined spacecraft. It was first deployed towards the Sun by the Centaur rocket before separation, while the spacecraft was in orbit around the Earth. Since for the first two years it is always pointing at the Sun, it cannot be used for communication with Earth until Cassini/Huygens is outside the Earth's orbit. It begins its main functions at Saturn, where, amongst other uses, it will pick up signals from the Huygens probe on its descent to Titan and relay them to Earth.

The spacecraft and their instruments

There are 27 teams involved in scientific investigations with Cassini and Huygens (Table 3); they have made 18 instruments. They have also formed themselves into nine Interdisciplinary Science Teams, which make no instruments but which bring together the scientific expertise to study certain science themes.

Participation in these teams – doing preparatory work, building gear, calibrating data – is how the team members earn the right to

Table 3. Science investigations.

Instrument	*To investigate*
	On Huygens
Aerosol Collector Pyrolyser (ACP)	Aerosol sampler and oven
Descent Imager and Spectral Radiometer (DISR)	Images and spectra
Doppler Wind Experiment (DWE)	Zonal winds, by tracking from Cassini
Gas Chromatograph and Mass Spectrometer (GCMS)	Chemical profile of Titan's atmosphere
Huygens Atmospheric Structure Instrument (HASI)	Physical profile of Titan's atmosphere
Surface Science Package (SSP)	State of Titan's surface
	On Cassini
	Optical Remote Sensing
Composite Infrared Spectrometer (CIRS)	Temperature and composition of surfaces and atmospheres within the Saturn system
Imaging Science Subsystems (ISS)	Multispectral imaging of Saturn, Titan, rings and the icy satellites to observe their properties
Ultraviolet Imaging Spectrograph (UVIS)	Spectra and low-resolution imaging of atmospheres, and imaging of atmospheres and rings for structure, chemistry and composition
Visual and Infrared Mapping Spectrometer (VIMS)	Spectral mapping to study composition and structure of surfaces, atmospheres and rings
	Microwave Remote Sensing
Cassini Radar (RADAR)	Radar imaging, altimetry and backscatter of Titan's surface
Radio Science Subsystem (RSS)	Atmospheric and ring structure, gravity fields and gravity waves
	Fields, Particles and Waves
Cassini Plasma Spectrometer (CAPS)	*In situ* study of plasma within and near Saturn's magnetic field
Cosmic Dust Analyzer (CDA)	*In situ* study of ice and dust grains in the Saturn system

Continued ►

Table 3. Science investigations – *cont.*

Instrument	*To investigate*
Ion and Neutral Mass Spectrometer (INMS)	*In situ* compositions of neutral and charged particles with Saturn's magnetosphere
Dual Technique Magnetometer (MAG)	Study of Saturn's magnetic field and interactions with the solar wind
Magnetospheric Imaging Instrument (MIMI)	Global magnetospheric imaging and *in situ* measurements of Saturn's magnetosphere and solar wind interactions
Radio and Plasma Wave Science (RPWS)	Study of plasma waves, radio emissions and dust in the Saturn system

Interdisciplinary Science for Titan

Aeronomy
Atmosphere/Surface Interactions
Organic Chemistry

Interdisciplinary Science for Saturn

Plasma Circulation and Magnetosphere-Ionosphere Coupling
Rings and Dust
The Plasma Environment in Saturn's Magnetosphere
Atmospheres
Satellites
Aeronomy and Solar Wind Interaction

analyse the scientific data from the spacecraft, satisfy their scientific curiosity, write the papers and get the glory. Each team is led by a Principal Investigator (PI) or Interdisciplinary Scientist (IDS) who is usually a professor at a university and who probably proposed the project to ESA or NASA in the first place. These team leaders have the responsibility for delivering each project. They organize the teams, cajole people into working together (even if they fall out), arrange finance, fit their project into the interfaces with the others, and sort out problems when things go wrong; they live with the project full-time for a decade and they earn their status, but no extra money. They work with their colleagues, with postgraduate students and with research assistants in the United States, France, the United Kingdom, Germany, Italy, Austria, Hungary, Spain, Scandinavia, the Czech Republic and the Netherlands, as well as the European Space Agency and the Jet Propulsion Laboratory in California – altogether, over two hundred scientists.

European scientists, one of them Professor David Southwood of Imperial College, London, lead as PIs two experiments in the Cassini Orbiter. US-led teams supplied two instrument packages for the Huygens probe, and American experts contribute to three others. European scientists participate in all the instruments on Cassini and Huygens, with Dr John Zarnecki of the University of Kent at Canterbury the PI for the Surface Science Package.

UK scientists have been involved in five instruments on Cassini and Huygens. The Cassini Plasma Spectrometer (CAPS) contains an electron spectrometer (ELS) provided by a European team led by the Mullard Space Science Laboratory (MSSL) of University College London, and including the Rutherford Appleton Laboratory (RAL). The University of Oxford built the cooler and focal plane assembly for the Composite Infrared Spectrometer (CIRS) to investigate the chemistry, thermal structure and dynamics of the atmospheres of Saturn and Titan. The instrument was built in collaboration with NASA's Goddard Space Flight Center. MAG is a dual magnetometer experiment forming part of the particles and fields package (MAPS) on Cassini, and was made at Imperial College. The Cosmic Dust Analyzer (CDA) will measure the physical and chemical make-up of small dust particles which Cassini encounters, both on the voyage to Saturn and at Saturn itself – including the ring material. RAL has a hardware involvement in this experiment which was calibrated at the University of Kent. The UK involvement in the Cassini Imaging Science Subsystem (CISS) is through software developed at the Cassini Imaging Centre to design the cameras, handle data and analyse images. Software is also being developed for ring, satellite and asteroid dynamics and to generate ephemerides for Saturn's satellites.

UK scientists have also been involved with instruments on the Huygens probe. The Surface Science Package (SSP) will perform direct measurements of the surface of Titan. This package of instruments will tell us how hard and flat the landing site is. If it lands in an ocean it will be able to measure the ocean's depth, density, and the speed at which sound and light travels through it. From these data, scientists will be able to establish the ocean's composition. This instrument was made at the University of Kent and RAL. HASI, the Huygens Atmospheric Structure Instrument, will measure the atmosphere of Titan as the probe descends slowly by parachute. It was built with University of Kent participation.

Including the industrial teams over 4,000 people have been involved in the Cassini/Huygens mission. A Europe-wide industrial team built the Huygens probe, with prime contractor being Aerospatiale in France. UK industry participation included Logica, who wrote the Flight Software, the Martin Baker Aircraft company, who made the Descent Subsystem, IGG, who procured parts, and Irvin, who supplied Parachutes. Including the ESA costs, the NASA costs and the costs in the national institutes which provided the instruments, Cassini/Huygens will cost the world between 0.5 and 1 billion dollars. Britain's share will be about £40M, funded by the Particle Physics and Astronomy Research Council, and most of it spent in British industry, buying equipment to technology-stretching specifications. For instance, the all-important parachute system (if it doesn't work, Huygens fails) relies on parachutes made of a fabric which must be specifically chosen so that it can be successfully stored for several years at very low temperatures and without emitting gases which affect the unusual instrumentation in the spacecraft; the parachutes must be quickly deployed by a mechanism which, however, does not break the delicate equipment in the probe; the support lines must not tangle even though the probe is spinning in its descent; and the parachutes have to be right to work in an atmosphere which is colder and denser but 1/7 the gravity compared with Earth's. You have really to understand parachutes to get all this right; and of course this rigour of understanding helps British industry make better parachutes for use on Earth.

The Cruise to Saturn

Huygens will remain dormant for nearly seven years, all the way from the Earth to Saturn, except for six-monthly checks. Cassini, on the other hand, will be busy. As it traverses the Solar System, astronomers will look for departures from its nominal trajectory caused by gravitational waves from distant moving masses (they will find these motions, if there are any that are large enough, by looking at the 'wow' in Cassini's radio transmissions). The Cosmic Dust Analyzer will measure the physical and chemical composition of small dust particles. This will be the first time that *in situ* chemical data on interplanetary dust, especially around the asteroid belt and Jupiter, have been obtained. The results will help to identify the origins of interplanetary dust grains whether they are from pulverized asteroids or melted comets, or whether they are primordial material from the origin of the Solar System, or interlopers from interstellar space (or, as is likely, all of these). As the spacecraft approaches Saturn, Cassini's instruments will become more and more active as they are calibrated, as they take long-range images for navigational and test purposes, and as their operators get in some practice for the tour around Saturn.

Hitting the target

In mid-2004, Cassini/Huygens will manoeuvre for the approach to Saturn. As it nears Saturn it will have been observing the planet and its satellites to pinpoint their positions. Terrestrial telescopes, particularly the Carlsberg Meridian Telescope, on La Palma, are being used to update the calculations of the satellites' orbits; but the positions of the satellites are known only to within 3000 km from these observations. Cassini has to know Titan's position to within a few tens of kilometres if it is to deposit Huygens in the right place, and the positions of the other satellites to the same accuracy in order to make low-pass orbits over their surfaces.

In June 2004, Cassini makes its closest approach to Phoebe, Saturn's most distant satellite. It won't get a second chance to see Phoebe so closely, because Cassini's four-year tour of the Saturn system does not bring it out this far again. Phoebe's orbit is inclined to Saturn's equatorial plane and it orbits the wrong way round (retrograde). Presumably it is a captured object, an asteroid perhaps. This will be the time to find out, as Cassini zooms past, its cameras trained on this unusual moon.

On July 1, 2004, Cassini/Huygens will manoeuvre into orbit around Saturn. The engineers will be busy: Cassini will be braking by firing its motors for 100 minutes to lose enough speed to go into orbit around Saturn. The scientists will be busy too: at about 20,000 km above its surface, this will be the closest Cassini gets to Saturn, so they want a good look, not only at Saturn but also at this region of space, to see what fields and particles it contains. They will get a really good look at the ring system, as Cassini goes through one of the gaps.

The Descent of Huygens to Titan

If everything goes according to plan, on November 6, 2004, Cassini will release the still dormant Huygens on course to Titan. As

Huygens leaves, Cassini will give it a spin so it is stabilized and drops towards Titan with its front end forwards. Cassini will continue on its trajectory, aiming the high-gain antenna towards Titan to receive the radio signal from Huygens. After a 22-day drop towards Titan, Huygens will be powered up, its sensors will come alive, and its brain will be activated, ready to deploy Huygens' mechanisms. The first mechanisms that have to work are those which stop Huygens from burning up like a meteorite in Titan's atmosphere. At about 500 km above its surface, where Titan's atmosphere begins, Huygens' accelerometers will sense the first slowing in its descent speed.

Huygens is built like a clam with a bivalve outer shell, which will act both as a brake and a thermal shield. The gas temperature in front of the shield will rise to 12,000°C. It is uncertain what chemical reactions this heat will provoke in Titan's atmosphere, and engineers have left a wide safety margin in the design of the insulating tiles of the front heat shield in case these reactions release extra heat. Three minutes after first encountering Titan's atmosphere, at an altitude of about 180 km, the deceleration will peak at about 16*g*, while the speed will drop to a Concorde-like 1400 km/h. Every structure of the space probe will be severely stressed, because Huygens, at this stage a 350 kg structure, will suffer *g*-forces equivalent to a weight of 5 tonnes.

A 2.5-m pilot parachute will be released at these supersonic speeds. It will yank off the Back Shield and pull out the 8-m main parachute (Figure 1). Within a minute, the speed will reduce to less than 250 km/h – say the speed of a fast train. After protecting the scientific instruments by keeping the atmosphere out, Huygens' shell must be popped off to let the atmosphere in so the instruments can get at it. The front shield will be released at a height of about 150 km, at which height the atmospheric temperature is expected to be about −120°C. The main parachute used for braking at high altitude will be cut away after just 15 min, and be replaced by a smaller one (3 m) which will deploy at about 120 km altitude.

Before the batteries run out, Huygens has to complete its descent. It will take two hours, perhaps with the probe drifting sideways in the winds. The atmospheric temperature will drop from −120 to −200°C at the base of Titan's stratosphere, perhaps 50 km above the surface. Here the haze may begin to clear enough to give Huygens' cameras their first glimpse of the surface of Titan, between clouds of methane. The cameras will scan the scene below, as Huygens spins under the parachute.

Figure 1. The Huygens probe is yanked from its heat shield as it slows in the thick, opaque methane atmosphere of Saturn's satellite Titan in October 2004, as envisaged by an artist. (Photo: ESA.)

It simply is not known what the surface of Titan is like (an artist's impression is shown in Figure 2). An ocean of mixed methane and ethane, perhaps decorated with luridly coloured organic ice floes? A marsh of methane/ethane puddles? A dry landscape with geysers spouting methane from underground springs? Volcanoes throwing out ammonia and water? For the last few hundred metres of the descent, a landing light will help the spectrographs to look for methane in the surface. An acoustic sounder will listen for echoes.

By the time Huygens knows what the surface is like, it will be too late to do anything about it. There will be a thud, a squelch or a

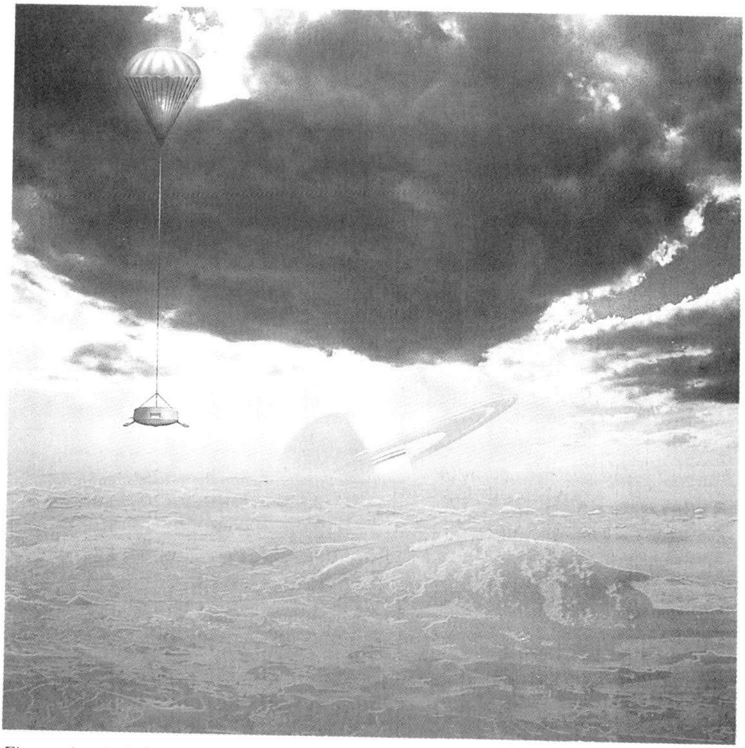

Figure 2. Artist's impression of Huygens lowering itself to Titan's surface, here envisaged as icy ground with fuming volcanic hillocks under lowering methane/ethane clouds and an atmosphere which, judging by the visibility of Saturn rising on the horizon, is considerably more transparent than actually expected. (Photo: ESA.)

splash as it hits the surface at a speed of about 30 km/h – the speed of a suburban car crash. Think of pushing the internal structure of Huygens, about the size and mass of a refrigerator, off the roof of a house. Is Huygens strong enough to survive? In May 1995, in an exciting test which the ESA engineers seem to have enjoyed, a full-scale model of Huygens was dropped with its parachutes from a balloon 37 km above Sweden. It successfully went through the planned procedure, and the instrument payload hit the Earth at a speed of 28 km/h. It survived – dented, but working (Figures 3 and 4).

If, as expected, Huygens survives the drop to Titan, the radar, chemical and imaging experiments will continue to operate, usefully but not for long, after the landing. The Surface Science Package, developed under British leadership, has its brief period of glory for the last few minutes of the mission. If the landing has been on a dry surface, a penetrometer will have gauged the hardness of the ground. A tiltmeter will show whether or not the surface is flat, and, if the surface is liquid, it might detect waves as Huygens bobs about. The acoustic sounder will probe the depth of the ocean or lake, while other instruments measure its density and the speeds of sound and light in the liquid.

Figure 3. A high-altitude balloon carried Structural Model 2 of Huygens to 37 km above Kiruna in Sweden in 1995 to test-drop it. (Photo: Fokker Space and Systems BV for ESA.)

Figure 4. Woolly-hatted engineers cast shadows as they contemplate the structural model of the Huygens probe sitting in the Swedish snow, intact and functioning, with its final parachute drooped out behind, after being dropped 37 km to simulate its descent to the surface of Titan. (Could it really have been this clean of snow after the impact at 9 m/s, or has it been spruced up for a photo opportunity?) (Photo: Fokker Space and Systems BV for ESA.)

If Huygens survives the heat, the cold, the wet and, above all, the impact, the time it will last on Titan's surface will be anything from 3 to 30 min; after 30 min the batteries run out. Because of the 80 min travel time of the radio waves from Saturn, by about the time John Zarnecki knows that the parachutes have worked, and Huygens has begun swinging from side to side as it floats down from Titan's sky, his Surface Science Package will have completed its job – half an hour's data collection at most, after a decade of making the equipment followed by a seven-year wait during the journey time.

Cassini explores the Saturn system

By this time, David Southwood's data-flow will just about be getting into its stride, and he will be taking time off from whatever high-powered job he has then, or retirement, to look at it. Cassini will tour the Saturn system from 2004 to 2008, measuring magnetic fields, plasmas and dust particle densities, and studying the meteorology of Saturn and the 'geology' of the satellites. The tour will consist of about 60 orbits, each one different from the next. They will have various orientations: some orbits will give edge-on views of Saturn's rings, some orbits will be over the poles of Saturn to view aurorae and other polar phenomena. The orbital periods will range from over a hundred days to less than ten. There will be various approaches to Saturn, from as close as three Saturn radii (about 180,000 km) to more than seven (420,000 km); and various approaches to Titan and the other satellites (including Mimas, Enceladus, Dione, Rhea and Iapetus).

There will not be much fuel for propellant left on board Cassini to control the tour, although small propulsive manoeuvres might be necessary. So every orbit must always return to Titan so that Cassini can use Titan as a gravity assist to position itself correctly for the next orbit. The large number of visits to Titan during the tour results in extensive coverage of this satellite. Fly-bys of the other satellites will be planned, if possible, to occur far enough from a Titan fly-by for supporting manoeuvres to be planned, the recorder emptied of previous data, and the ground system rested for a new and busy sequence. There will not be many fly-bys of the icy satellites, because they are too small to provide appreciable gravity assist, and to visit them costs propellant.

Mission extension

What after that? Cassini has the potential to go on beyond the four years. What happens depends on what it finds out, how much more there is to do, what the competing claims are for space scientists' attention (and there will be many other exciting things going on about 2009). But astronomers have already thought of a few possibilities to extend the Cassini mission:

Fly closer to Titan. Cassini could risk a crash landing on a close approach to Titan in order to examine its atmospheric characteristics at lower altitudes.

Fly closer to Saturn. After the four-year tour, Cassini will most likely be in a polar orbit and could risk damage by ring particles on a close approach to Saturn, inside the G Ring.

Investigate Saturn's rings more closely. There are some passages of moons near or through part of the rings in the few years following 2008, and Cassini could be positioned for a close look at how a small moon might plough through a ring.

More of the same. More time means being able to look at interesting things that have been found already, to greater accuracy. Not so dramatic as the above three, but maybe more useful.

Escape Saturn's gravity. This would be to investigate reaches of space more distant from Saturn.

Conclusion

There are not as many planetary scientists in Britain as, say, in France or Germany, and those British astronomers who study how the planets work are dispersed in relatively small groups at several universities and institutes. But these groups form a European virtual centre of excellence which has played a significant role in the Cassini/Huygens mission. The British space industry too – a British success story, with 6500 employees and a turnover of £500 million per year, apparently little known to the British public – is also a key part of the international team that built the Huygens probe. Working with American and other European partners, talented British men and women in industry and academia have used their skills and expertise in a grand global enterprise, of the sort that human beings do best, to create a machine in order to make a fundamental investigation of two of the Earth's distant yet closest neighbours. Late in 1997, I suspect that nearly all the 4000 people who worked on Cassini/Huygens will have paused in whatever they have turned to since their work on this mission and will have listened anxiously for news of the launch. I hope, as I conclude my article early in the year, that they will have heard good news and that, in a few years' time, they will be able to read about our better understanding of what happens on Saturn and its moon Titan, and remember that they contributed to these advances in knowledge.

Beyond Neptune – The Edge of the Solar System

ANDREW J. HOLLIS

The outer regions of the Solar System have attracted much interest in recent years. Firstly, four space probes (Pioneers 10 and 11, and Voyagers 1 and 2) have been tracked into the region beyond Neptune, and more recently several small (less than 500 km in diameter) objects have been detected. Perhaps as significant as any new sources of information has been the discovery of dust clouds (suggesting the existence of planetary systems of some sort) around several nearby main sequence stars.

The known size of the Solar System was doubled with William Herschel's discovery of Uranus in 1781, and another great leap came in 1846 when Neptune was found. Pluto, revealed in 1930, was for many years considered to be a large planet with a mass several times that of the Earth. However, after the discovery of its satellite, Charon, in 1978 it was calculated that the mass of the Pluto/Charon system was about 1/50th that of the Earth – much too small to have any significant gravitational influence on other outer planets.

During their long voyages, the paths of the deep space probes have been tracked carefully to see if they have been subject to gravitational perturbation by any unknown major body. No significant deviation has been noted, and it is probable that no large object accompanies the Sun beyond the orbit of Neptune. No major planet, black hole, brown dwarf star or other exotic body is likely to be found.

In 1943, a paper was published in the *Journal of the British Astronomical Association* by Kenneth Edgeworth, outlining a theory of the evolution of the Solar System. In his theory, small particles formed into a rotating disk around the primitive Sun. In the inner Solar System these combined to form the major and minor planets. However, beyond the orbit of Neptune the particle density was not high enough for a single planet, and many smaller bodies formed instead. Edgeworth then proposed that 'From time to time a member of this swarm of potential comets wanders from its own sphere and appears as an occasional visitor to the inner regions of the Solar System.' This explains the appearance of short-period comets. He proposed that comets were 'astronomical heaps of gravel without any cohesion' – a view proposed by R. A. Lyttleton with his 'sand-bank' model of comets. As this model is no longer considered feasible, this could explain why Edgeworth's theory was overlooked until relatively recently.

In 1950 the Dutch astronomer Jan Oort suggested that comets formed in the asteroid belt. Most were removed from this area as their orbits were perturbed by Jupiter, and formed a halo well beyond the planets in interstellar space. Perturbations by passing stars could explain the sudden appearance of bright comets in very long-period orbits. In 1951 Gerard Kuiper suggested that the icy composition of comets could be explained if they had formed in the region just beyond Neptune. Stellar perturbations can also remove bodies from the planetary region to this distant reservoir, and this could be considered as a development (perhaps independently) of Edgeworth's idea.

There are two distinct distributions of comets. Historically, orbit determination was relatively imprecise and only accurate for periods less than about 200 years, so comets with periods less than this are called short-period comets. Modern techniques are better and this limit should perhaps be increased. Comets with periods greater than 200 years are termed long-period.

Most of the short-period comets known follow low-inclination orbits that are prograde – in other words, they orbit the Sun in the same direction as the major planets, but are in more eccentric orbits. There are exceptions, the most notable of which is Halley's Comet, inclined nearly at 163° to the planetary orbits and usually well out of their plane. Comets in such orbits could form a higher proportion of the total, as undoubtedly many faint comets pass by the Earth undetected – comets can be seen only if the observing geometry is favourable. (An asteroid found in 1991, 5335 Damocles, follows a cometary shaped orbit but shows no evidence of cometary activity; it is either a solid rocky object or else a comet that is now defunct.) It is thought that the reservoir supplying these comets is at a distance of perhaps 30 to 100 AU from the Sun. This has now been called the Kuiper Belt.

Long-period comets appear to be evenly distributed in inclination. In other words, they can come from any direction to pass through the inner Solar System. This suggests that there is a

Table 1. Trans-Neptunian objects discovered up to August 1996.

Designation	*Perihelion distance (AU)*	*Aphelion distance (AU)*	*Incli-nation* °	*Apparent magnitude*	*Approximate diameter (km)*
1992QB$_1$	40.879	47.276	2	22.8	283
1993FW	41.570	45.889	8	22.8	286
1993RO	31.486	47.383	4	23.2	139
1993RP	34.863	43.795	3	24.5	96
1993SB	26.887	51.754	2	22.7	188
1993SC	32.154	47.307	5	21.7	319
1994ES$_2$	40.352	51.020	1	24.3	159
1994EV$_3$	41.209	44.701	2	23.3	267
1994GV$_9$	41.077	46.258	0.1	23.1	264
1994JS	32.674	53.030	14	22.4	263
1994JV	35.251	35.251	17	22.4	254
1994JQ$_1$	42.983	45.176	4	22.4	382
1994JR$_1$	34.745	44.846	4	22.9	238
1994TB	27.308	51.189	12	21.5	258
1994TG	42.254	42.254	7	23.0	232
1994TH	40.940	40.940	16	23.0	217
1994TG$_2$	42.448	42.448	2	24.0	141
1994VK$_8$	39.316	46.472	1	22.0	389
1995DA$_2$	32.617	40.158	7	23.0	169
1995DB$_2$	40.152	52.401	4	22.5	266
1995DC$_2$	42.000	45.547	2	22.5	338
1995FB$_{21}$	42.426	42.426	1	23.5	169
1995GJ	39.006	46.808	23	22.5	301
1995GA$_7$	34.751	44.160	4	23.0	202
1995GY$_7$	41.347	41.347	0.9	23.5	165
1995HM$_5$	29.796	48.894	5	23.1	161
1995KJ$_1$	43.468	43.468	3	22.5	361
1995KK$_1$	31.981	46.969	9	23.0	166
1995QY$_9$	29.489	49.468	5	21.5	260
1995QZ$_9$	34.639	44.294	20	22.5	235
1995WY$_2$	47.551	47.551	10	23.4	290
1995YY$_3$	30.665	48.122	0.5	23.4	145
1996KV$_1$	41.292	41.292	10	22.9	268
1996KW$_1$	46.602	46.602	6	23.4	281
1996KX$_1$	35.704	43.381	1.5	23.9	131
1996KY$_1$	35.712	43.322	31	23.3	126

reservoir of cometary nuclei forming a 'shell' round the Solar System, beyond the Kuiper Belt. This is the Oort Cloud.

Discovery of distant bodies

Pluto was discovered by Clyde Tombaugh (who died in January 1997). He carried out a photographic search of the entire sky north of declination $-30°$ to a limiting magnitude of 16. This magnificent record still exists, and is used to help refine the elements of newly discovered bright asteroids when images can be found on Tombaugh's plates. Following the discovery, he extended the survey to magnitude 17.5 for regions near the ecliptic, but with no further success. Charles Kowal carried out a programme to search for slow-moving Solar System bodies, and in 1977 discovered asteroid 2060 Chiron, the first (excluding Pluto and comets) of the outer Solar System objects. Initially, Chiron was thought to be an inert object but in 1988 (on its approach to perihelion) a coma was detected, suggesting that it may have a cometary composition. Chiron has an orbit that lies between those of Saturn and Uranus. At the time of writing five other objects have since been found with perihelion distances close to Saturn's orbit – three of these have orbits with aphelia further from the Sun than Neptune. These bodies are known collectively as the Centaurs.

The first true trans-Neptunian was discovered in 1992 by David Jewitt and Jane Luu at 41 AU from the Sun, and is designated 1992 QB_1. The TNOs (trans-Neptunian objects)[1] can be separated into two groups. The first have essentially circular orbits with a radius of 40 to 46 AU, and the second have an appreciable eccentricity, giving a perihelion distance near that of Neptune – these orbits are similar in shape to the orbit of Pluto, and like Pluto they appear to be in a 3:2 resonance with Neptune, which means that for every three orbital revolutions of Neptune they complete two. They have thus been christened 'Plutinos'. By early August 1996, 36 TNOs had been discovered and 14 of these are Plutinos. Most of their orbits are of low inclination, though this may be a biased result of the search pattern. The maximum value is 31° for 1996KY_1, but only seven examples have orbits inclined to the ecliptic by greater than 10°.

The method of searching is to image a field nearly at opposition and very near the ecliptic. The typical field of CCD imagers is of the order of 7′ or 8′ wide with a 2 to 3-m aperture telescope. It would be unusual to find a body in a high-inclination orbit, as it would have to be in this small field at precisely the right time.

Estimates of the size of these bodies have been made by assuming that they have an albedo typical of the comets, at 0.04. This is not unreasonable as they appear to show similar spectra to comets. The numbers found in the small fields suggest that there are perhaps 35,000 bodies of diameter 100 km and above in the zone 30–100 AU from the Sun, which is several hundred times the number of asteroids of that size in the main belt. The largest known is approximately 380 km in diameter, but as the search area has been very small much larger examples probably remain to be found. This result contrasts with only about 230 objects larger than 100 km diameter in the main asteroid belt. From the Earth, it is likely that objects 50 km or larger could be detected; however, smaller ones would be very difficult to spot unless a fully computerized search programme is introduced. Past observational searches would probably have detected any large members of the Kuiper Belt, and it is unlikely that examples much larger than 600 km diameter exist.

Of the 36 TNOs found by June 1996, the brightest is of magnitude 21.5 and the faintest is magnitude 24.5 – very near the limit of detectability for ground-based telescopes. Current searches are carried out during the spring and autumn. In the winter and summer skies the Milky Way crosses the ecliptic, making it difficult to identify the slow-moving objects because of the large number of faint stars. During the spring of 1995, images were taken using the Hubble Space Telescope's Wide Field Planetary Camera. By combining 34 of these images (only 4′ in size), Anita Cochran and her team removed stars and galaxies from the field. They were left with hundreds of faint images near the magnitude-28 limit of the telescope. Many of these were spurious noise, some arising from cosmic-ray events. However, they were able to predict the typical motion of a body at large solar distance, and from this decide that about 50 of the images were probably real. This result implies that there are hundreds of thousands of objects of about 20-km in diameter in the Kuiper Belt. This is the size of a typical cometary nucleus, and reinforces the belief that the Kuiper Belt is the main source of short-period comets.

There does not seem to be an overabundance of smaller bodies, so we may assume that either they have not been discovered as yet or that the Kuiper Belt is composed of the products of accretion rather than fragmentation (as seems to be the case for the main asteroid belt). The objects' precise composition is unknown. Pluto has a high albedo, suggesting an icy composition, perhaps similar to that of Triton (Neptune's large satellite). A tenuous atmosphere formed recently as Pluto approached and passed perihelion, and was detected during the occultation of a star. The Kuiper Belt objects appear to have a low albedo (approximately 0.04) and are probably of cometary composition. Further research is needed to confirm this.

Evolution

The existence of circumstellar planetary disks seems to confirm theoretical studies of the formation of planetary systems. Dust from the inner Solar System is gradually cleared away. Much is swept up by the planets, and they are continually increasing in size. This increase is very small relative to their size. The Earth, for example, collects some 15,000 tonnes of dust every year, but this accumulation goes almost completely undetected, except when bright meteors are seen. In addition, the drag of radiation pressure from the Sun causes small particles to spiral inwards to it (this is called the Poynting–Robertson effect).

Planets are known to grow by accretion. Planetesimals in similar orbits collide and coalesce, and this process continues until all available bodies have been 'swept up'. In the inner parts of the Solar System this process happened rapidly as orbital periods are quite short – only 4 or 5 years at the distance of the asteroid belt. The outer planets move so slowly that their rate of accretion is low. Theoretical studies indicate that the time it would take for a planet beyond Neptune to grow to ten thousand kilometres or so is greater than the age of the Solar System. Significant quantities of dust and small planetesimals should therefore be present. The dust disks around other stars seem to confirm this depletion in inner regions.

A simple illustration of why planet growth is so slow in the outer Solar System can be made. Pluto orbits the Sun in a period of just less than 250 years. It moves some 25 times slower than an asteroid in the main belt. All things being equal, it would collide with other planetesimals 25 times less often than if it were in the main belt. Simplistically, it would take 25 times longer to grow by accretion. This implies that if asteroid growth took 150 million years in the main belt, it would require the age of the Solar System to complete at Pluto's distance. This very simple scenario does not take into account the presence of large planets like Uranus and Neptune, or the role of gravitational disturbance by Jupiter in controlling the distribution of planetary size, but it does illustrate the point.

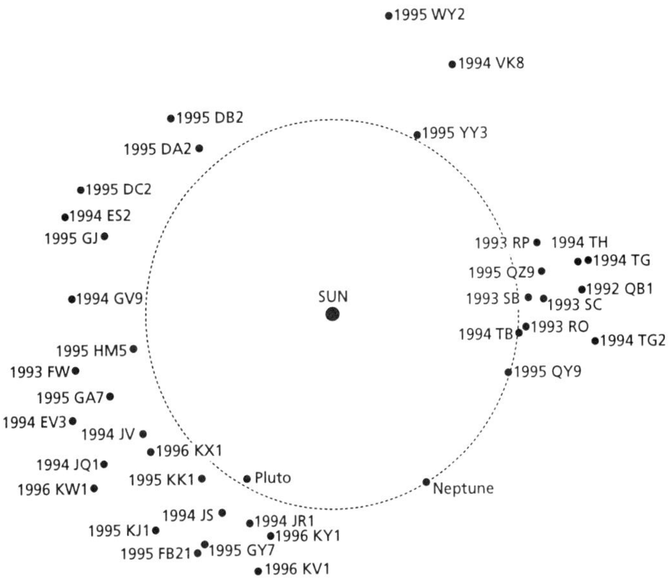

Figure 1. Location of Kuiper Bett Objects and Neptune's orbit as known in mid-1996.

It is unlikely that direct observational evidence of the structure of the Sun's circumstellar disk will be possible to the same degree as for those we can see around nearby stars. The main problem is that we are too close to be able to see the whole picture. This is a similar problem to that of determining the shape of our Galaxy. However, it seems probable that what we can see of other stars is likely to be typical of our own. It is a two-way exercise, in that we are not able to image planetary bodies in other stellar systems (though some larger ones have been inferred from astrometric measurements), so that theoretical deductions may be drawn from a combination of the two sets of studies.

Conclusion

There are only a small number of teams carrying out observations to discover distant Solar System objects, and only a limited time available for the Hubble Space Telescope to take images. It is unlikely that a search to discover every large member of the Kuiper

Belt will be made, and indeed it is probably not going to be a cost-effective use of time and resources. Having said that, the first four years have revealed 36 examples, whereas the fifth asteroid in the main belt was not discovered until 45 years after the first one. The empty spaces of the Solar System are being filled with new breeds of object, and the next main area for a new search would seem to be the region beyond the Kuiper Belt, where the circumsolar disk may be found. This is likely to be a region within the Oort Cloud extending to perhaps 2000 AU from the Sun.

Each time a new area is studied with new instrumentation, something unexpected turns up. This trend seems to be continuing at an increasing rate, and it seems probable that it will be well into the next century before we can be confident that we have at last come to grips with the whole picture of our own backyard. However, whenever we come to grips with any such problems new information usually comes to light. The history and the present (to say nothing of the future) of our Solar System will undoubtedly hold a few more surprises in the years to come.

Note

1. At present, the terminology is still rather fluid. 'Trans-Neptunian object' and 'Kuiper Belt object' are two names for the same thing, and 'Edgeworth–Kuiper object' is another term that is being used. Astronomers will presumably settle on a single term in due course.

The Enigma of Tektites

JOE McCALL

In August 1996, the media turned their attention to meteorites, following the reported possible discovery of evidence of life on Mars in a meteorite from Antarctica. In subsequent television programmes, scientists were shown picking up small glassy objects known as tektites in the desert of South Australia. At that time, I had already commenced a literature search and review of the state of knowledge and the still viable hypotheses concerning tektites. Fortuitously, in August I was in China when the news of the Mars controversy broke, and was surprised to find tektites from Guangdong Province in Southern China displayed at the exhibition accompanying the 30th International Geological Congress. Although I was not able to secure a dinosaur's egg (several, including a nest, were on display!), I was able to negotiate to bring away five tektites from Guangdong. This collection, from a little-known area where tektites are found, augmented my literature search.

Historical

The existence of tektites has been known for centuries. The first reports of them are from ancient China (Barnes 1969; Xu Dao-Yi *et al.* 1989; Wenzhli Lin *et al.* 1995; Figure 1a), where they were known as *lei-gong-mo*, which means 'ink sticks thrown and spread from the sky by the thunder god', or *lei-gong-shih*, which means 'stool of the thunder god'. The earliest record is in the 'Records of Heterodoxy Outside Nanling', according to which tektites were found on the Leichou Peninsula by Liu Xun, an official of the Tang Dynasty, in the middle of the tenth century AD. They were later described from Czechoslovakia by Josef Mayer in 1788 and from Australia by Charles Darwin (1844, Figure 2b), who believed them to be volcanic bombs. They were reported from Indochina by A. Lacroix (1932) and the Ivory Coast, also by Lacroix (1934), and lastly from Texas by H. B. Stensel in 1936 (Barnes 1940).

For a long time the popular view was that tektites are extra-terrestrial visitors, a sort of meteorite (and a few scientists still

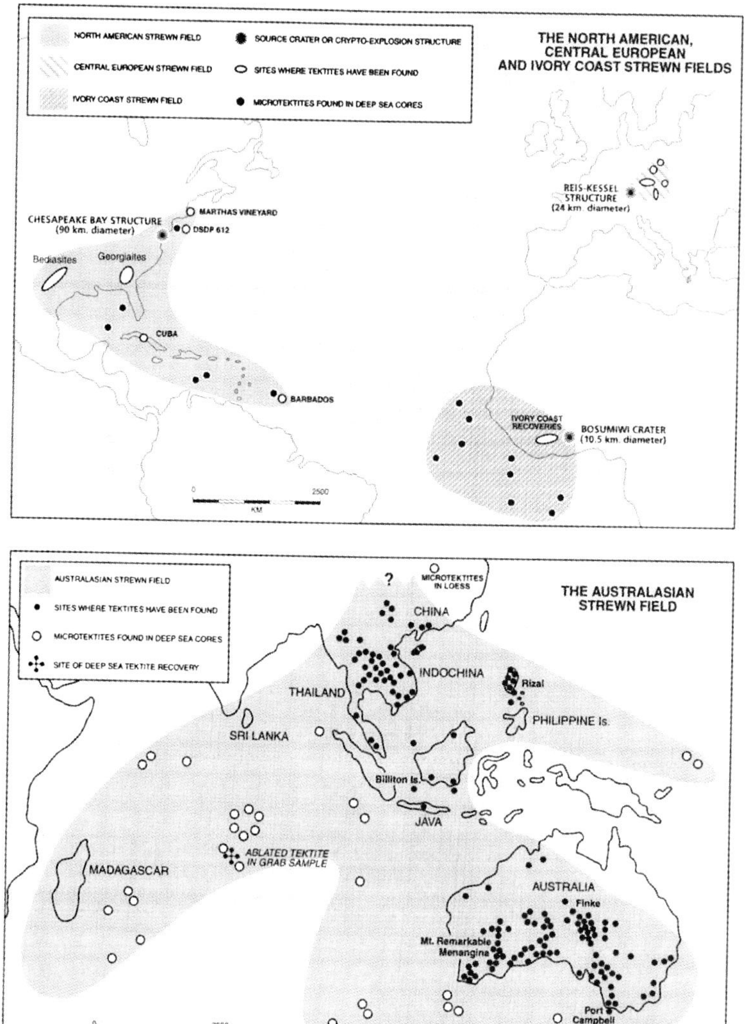

Figure 1. The four strewn fields: (a) The Australasian strewn field, and (b) The North American, Central European and West African strewn fields.

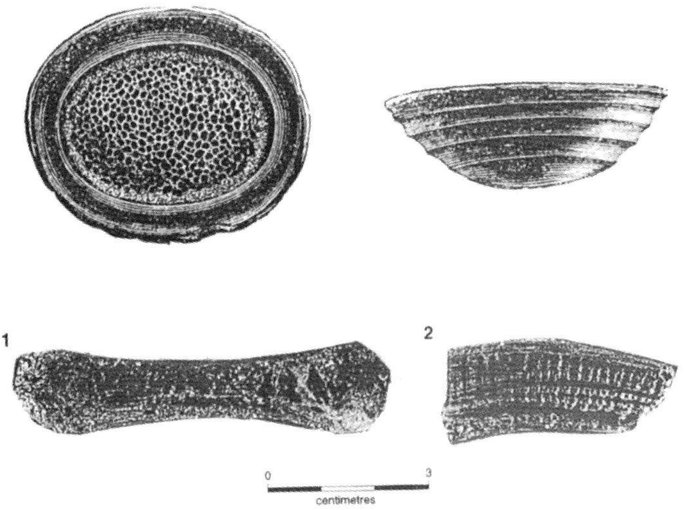

Figure 2. (a) Charles Darwin's drawing of a flanged button australite which he believed to be a volcanic bomb (from Darwin 1844). (b) Tektites from Hainan Island, South China, from Xu Dao-Yi et al. *(1989): (1) dumbell-shaped splash form; (2) layered tektite glass.*

adhere to such a hypothesis). However, analysis of lunar rocks brought back by the Apollo missions of 1969–72 showed that they certainly did not come from our satellite (Schnetzler 1971, Taylor 1973). Since then, researches of immense thoroughness, technical complexity and ingenuity have shown that the material of these glassy bodies must be terrestrial, even if the tektites have been ablated in re-entry to the atmosphere. The most widely accepted view is that they were blasted out of the continental subsurface zone a few hundred metres down by large-scale asteroidal or cometary impacts, and that the source material was sedimentary rock. Yet the enigma of tektites is far from resolved. The aim here is to summarize briefly what is at present known about tektites and to list the unanswered questions.

The strewn fields

The term *tektite* is best restricted to small, pebble-like, naturally occurring objects made of siliceous glass, pale green to black in colour, which occur in distinct geographical *strewn fields*. It is best

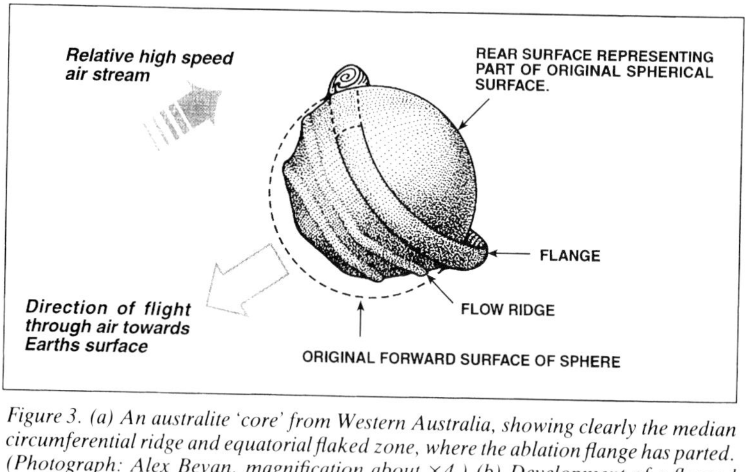

Figure 3. (a) An australite 'core' from Western Australia, showing clearly the median circumferential ridge and equatorial flaked zone, where the ablation flange has parted. (Photograph: Alex Bevan, magnification about ×4.) (b) Development of a flanged button australite by ablation. (From McNamara & Bevan 1991.)

to exclude from the classification microscopic glassy spherules and fragments that occur, for instance, in Haiti, at the Cretaceous-Tertiary boundary, and impactite glasses associated with impact craters (Taylor & Koeberl 1994). There are four known strewn fields: the North American, Central European, Ivory Coast and Australasian (Figure 1). The age of formation of the tektites, when they solidified from a molten state, can be determined radiometrically using the K-Ar and Ar-Ar decay ratios, and fission track measurements. Thus we know that the North American strewn field is the oldest (34–35 million years), followed by the Central European (about 14.7 million years) and the Ivory Coast (about 1 million years), with the immense Australasian strewn field the youngest (0.78 million years). However, all tektites have been transported by geological processes and eroded since their arrival on the ground, and this gives rise to what is known as the 'age paradox' – they are usually found in younger geological formations than their radiometric age indicates. Thus the tektites in Texas and Georgia are found in Oligocene and Miocene formations respectively, and those found in Victoria occur within slightly older Holocene sediments than those found in Western Australia. Recently, however, tektite fragments have been found in sediments in Barbados of late Eocene age, equivalent to their radiometric age, so the age paradox is no longer of great significance.

Enormous numbers of tektites have been collected, more than 2000 from North America, more than 55,000 from Central Europe, more than 2000 from the Ivory Coast, about 600,000 from Southeast Asia and about 100,000 from Australia (McNamara & Bevan 1991). In Western Australia they tend to be shed on to the surface of present-day clay pans, and can be picked up quite easily by walking across the surface with sunlight from a certain direction, so that they glisten strongly.

There are three essential types of tektites: *splash forms*, *ablated splash forms* and *Muong Nong* (*layered*, *block*, *irregular*) (Koeberl 1994). However, it has become apparent that there are intermediate types between the second and third types, layered but with splash forms (Izokh & An 1983). The splash forms (spheres, disks, teardrops, peardrops, dumbells, boat shapes, etc.) were projected from the Earth as molten objects and then solidified. Many (perhaps all) of them were later melted again peripherally by ablation, it is believed on re-entry into the Earth's atmosphere. Ablation was a secondary process, imposed on splash forms. The most perfect secondary ablation forms of the splash forms are the flanged buttons (Figures 3b, 4, 5a). Chapman & Larson (1963) carried out spectacular wind-tunnel experiments with gelatine and tektite glass, reproducing the flanged button form perfectly (for example McCall

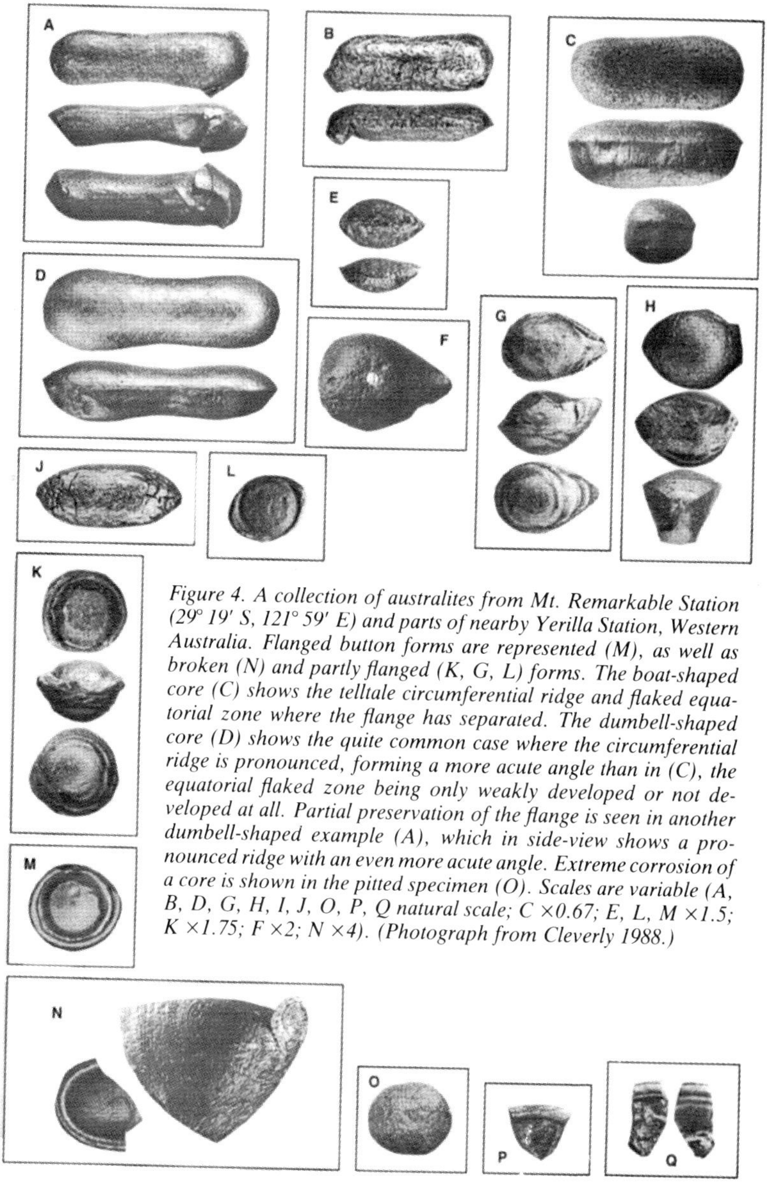

Figure 4. A collection of australites from Mt. Remarkable Station (29° 19′ S, 121° 59′ E) and parts of nearby Yerilla Station, Western Australia. Flanged button forms are represented (M), as well as broken (N) and partly flanged (K, G, L) forms. The boat-shaped core (C) shows the telltale circumferential ridge and flaked equatorial zone where the flange has separated. The dumbell-shaped core (D) shows the quite common case where the circumferential ridge is pronounced, forming a more acute angle than in (C), the equatorial flaked zone being only weakly developed or not developed at all. Partial preservation of the flange is seen in another dumbell-shaped example (A), which in side-view shows a pronounced ridge with an even more acute angle. Extreme corrosion of a core is shown in the pitted specimen (O). Scales are variable (A, B, D, G, H, I, J, O, P, Q natural scale; C ×0.67; E, L, M ×1.5; K ×1.75; F ×2; N ×4). (Photograph from Cleverly 1988.)

1973, Plate 80). Unfortunately, almost all the flanged buttons come from Australia; there is a wide disparity in climatic and vegetational conditions between Australia and South-east Asia, and this may be part of the reason for the greater degradation of South-east Asian tektites of the same radiometric age of formation. None the less, careful study of Australian tektites reveals the scar left behind by separation of the ablation flange (Figures 3a, 4, 5b) and careful study of tektites from South-east Asia and China reveals that the same scars are faintly visible (Figure 6), and they can even be detected on some older North American and Central European tektite specimens. So it is certain that some tektites from these three strewn fields suffered ablation, though it is by no means certain that all did.

The picture becomes complicated by the Muong Nong forms. Many of these are blocky and very large – up to at least 12.8 kg (McNamara & Bevan 1991). It has been a popular belief that they were produced by multiple impacts in 'puddles', suffering some flow on the ground after arrival and sintering into large masses (for example, Wasson 1991). However, close examination in the field in north-east Thailand lends no support to this hypothesis (Fiske *et al.* 1996), which is in any case theoretically unlikely and also unlikely considering the chemical uniformity of tektites. The large irregular masses seem rather to be blocks broken off even larger masses, and there is no evidence at all of sintering.

Whatever rationale is developed for tektites must accommodate the Muong Nong type, which has been recorded in Indochina, the Philippines and China over an extent of more than 2000 km, and again in the North American strewn field (Texas, Georgia, DSDP Site 612 off New Jersey) over a similar distance. As some layered forms display ablation surfaces (for example those from South China) and there are intermediate types with splash form (Izokh & An 1983), it appears that these layered types must have been projected upwards from a single impact site, and it many cases adopted splash forms, many of them at least being subsequently ablated on re-entry into the atmosphere. They must have been widely dispersed. The two cases of such forms spread over distances of 2000 km cannot surely be otherwise explained.

Muong Nong tektites display certain compositional differences from other tektites (slightly more water and volatiles, traces of minerals such as chromite, zircon, quartz, corundum, rutile and monazite, the zircons being commonly shocked at high temperatures).

Figure 5. Top right: Four views of a small ablated tektite obtained from a grab sample in the Indian Ocean at 12° 37′ S, 78° 30 ′ E at a water depth of 5300 m. This is the first ablated tektite to be recovered from sediments on the ocean floor. The flattened disk form and size are almost identical to that of the small Nallah meteorite, a bronzite chondrite from Western Australia (McCall and Cleverly 1969). (Photograph supplied by B. P. Glass, reproduced from Glass et al. *1996.)*

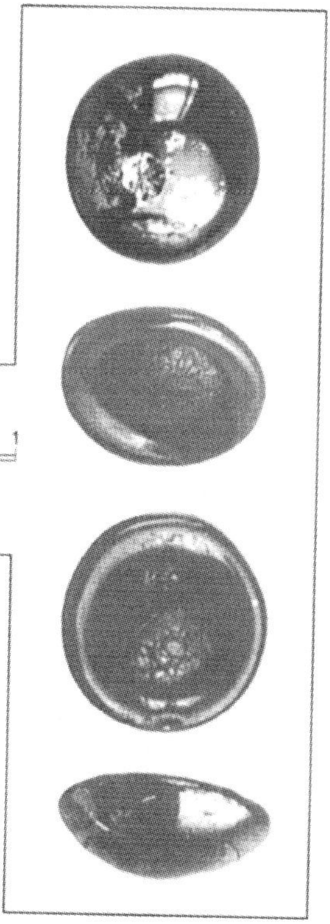

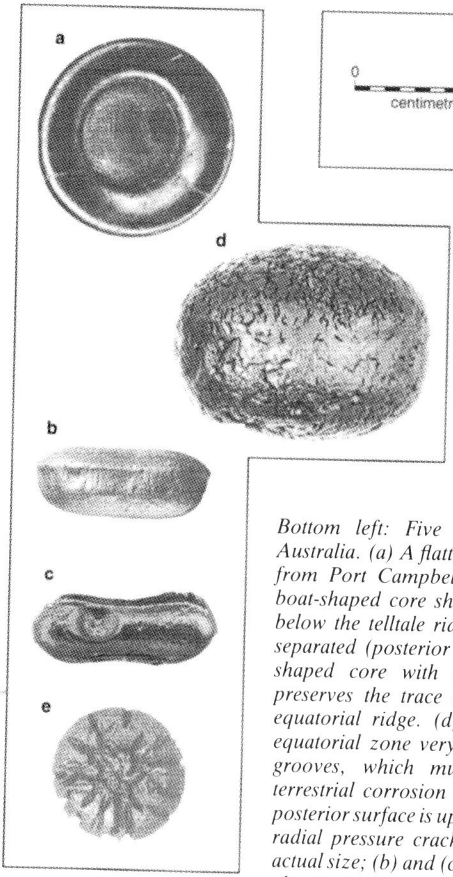

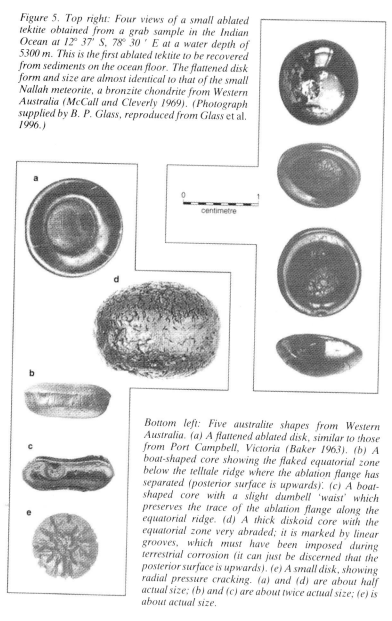

Bottom left: Five australite shapes from Western Australia. (a) A flattened ablated disk, similar to those from Port Campbell, Victoria (Baker 1963). (b) A boat-shaped core showing the flaked equatorial zone below the telltale ridge where the ablation flange has separated (posterior surface is upwards). (c) A boat-shaped core with a slight dumbell 'waist' which preserves the trace of the ablation flange along the equatorial ridge. (d) A thick diskoid core with the equatorial zone very abraded; it is marked by linear grooves, which must have been imposed during terrestrial corrosion (it can just be discerned that the posterior surface is upwards). (e) A small disk, showing radial pressure cracking. (a) and (d) are about half actual size; (b) and (c) are about twice actual size; (e) is about actual size.

The conclusion of both Schnetzler (1992) and Koeberl (1992) was that layered forms (Figure 2b) were less homogenized and were derived from a sedimentary source, not soil or loess, within a few hundred metres of the surface. The nature of the mineral inclusions certainly supports this. However, no residual fragment of source rock material has ever been found within a tektite.

Tektites geochemistry

Tektites are highly silicic, comparable with rhyolitic igneous rocks, but are more consistent in composition and have a minute water content, unlike rhyolites (Schnetzler & Pinson 1963). The trace element (including rare earth elements) and isotopic contents show that the source is terrestrial, the $^{26}Al:^{10}Be$ ratios being quite conclusive (Taylor & Koeberl 1994). The water content, though extremely low, is greater than in the dry lunar rocks (Koeberl & Beran 1988), and the lack of silicic rocks on the lunar surface would seem also to preclude a lunar origin. A very low water content is quite compatible with observations in nuclear explosion cratering (Taylor & Koeberl 1994). There is nothing in the chemistry to support a lunar or terrestrial igneous origin, but it does support a terrestrial sedimentary rock source.

Microtektites

Great advances have been made in the last few years in recovering and studying *microtektites*, microscopic glass bodies, from deep-sea cores (for example Koeberl & Glass 1988; Glass 1989; Glass *et al.* 1991; Glass 1993; Glass & Wu 1993; Glass & Pizzuto 1994). Despite their reservations about loose usage of the term 'tektite', Taylor & Koeberl (1994) accept that microtektites recovered from deep-sea cores are closely related to the tektite strewn fields – this is apparent from the distributions shown in Figures 1a and 1b. Radiometric dating confirms this, confirming their age of formation; and, in the case of the West African offshore occurrences, their stratigraphic position in relation to the Jaramillo geomagnetic event also confirms the relation to the Ivory Coast tektites (Glass *et al.* 1991). In DSDP Site 612, off the New Jersey Coast, tektite fragments are found associated with microtektites, and here there is no 'age paradox' as the stratigraphic level is around 35 million years (late Eocene) (Koeberl & Glass 1988; Glass 1989). Tektite fragments and microtektites have also been found at this level onshore in

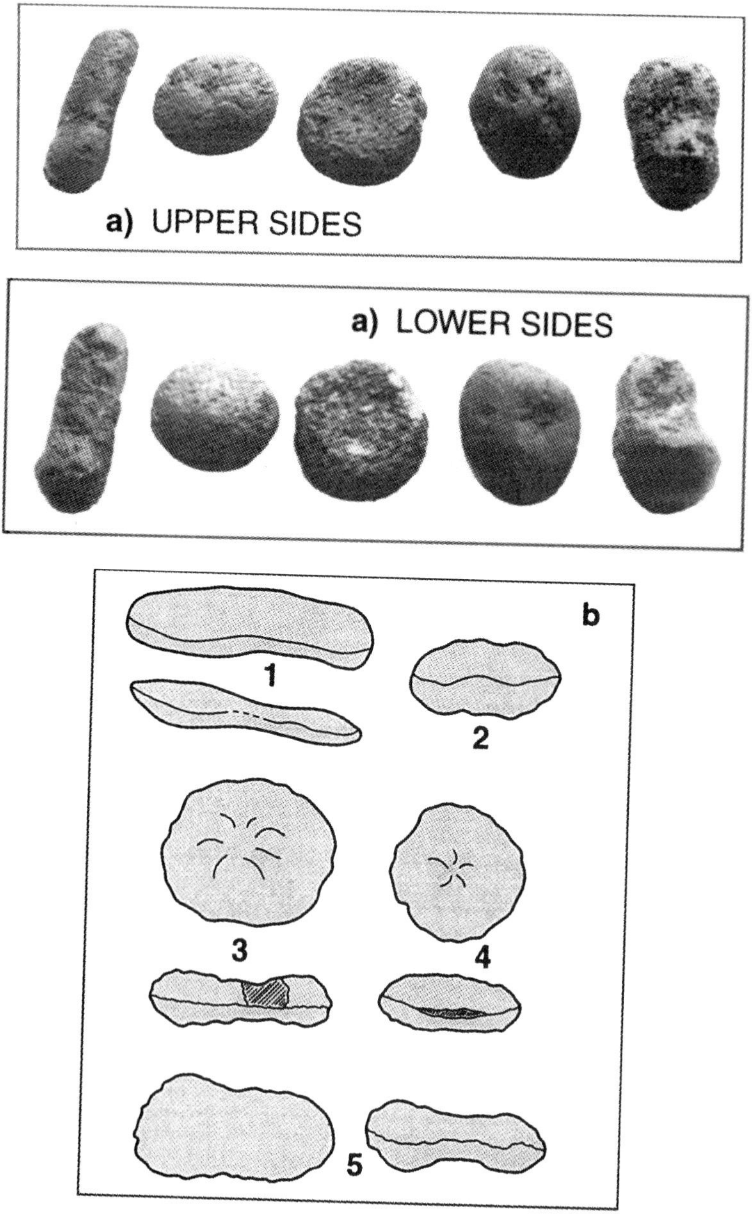
a) UPPER SIDES
a) LOWER SIDES
b
1
2
3
4
5

Barbados (Saunders *et al.* 1984; Sanfilippo *et al.* 1985; Koeberl & Glass 1988). Microtektites extend the limits of the strewn fields vastly (Figures 1a, 1b). Recently an ablated tektite has been recovered from a grab sample in the Indian Ocean (Figures 1a, 5) (Glass *et al.* 1996), confirming the extension of the Australasian strewn field already indicated by microtektite recoveries. Strictly, the geochemistry of microtektites is slightly different from that of tektites, but this may be due to different methods of analysis (Koeberl 1994).

Possible source structures

Of the four strewn fields, three have been related to possible source craters and cryptoexplosion structures.

Rieskessel, Germany

The Ries basin in southern Germany (Figure 1b), with a centre at 48° 53′ N, 10° 37′ E, is a depression 24 km in diameter. It is associated with a smaller structure, the Steinheim Basin, 3.5 km in diameter and 30 km to the south-west. The Ries has been dated at 14.8 (±0.7) million years by the K:Ar ratio method and 14.0 (±0.6) million years by the fission track method. Both structures are associated with shatter cones and shock deformation of quartz, and the Ries structure is associated with the high-pressure polymorphs of silica, coesite and stishovite. Horz (1982) reviewed the various studies of the Ries: different types of ejecta can be linked there with

Figure 6. The five tektites collected recently by the author from the Chinese mainland, Guangdong Province.

(a) Photographs of the upper and lower sides: note the strong contrast between the upper and lower (anterior and posterior?) surfaces in the elongated boat-shaped form (left); the disk (second from left); and the doughnut-shaped form (centre), which has a coarsely scoriacious character. In the 'boat' (length 9.3 cm, weight 71.11 g) the posterior surface (? above in the lower view) is much more coarsely pitted; the 'doughnut' (maximum diameter 6.8 cm, weight 104.38 g) has one coarsely pitted surface and one finely pitted; the 'diskoid' has one surface with small rounded pits and one with larger, irregular and linear pits.

(b) Drawings of the five specimens: (1) Elongated boat-shaped form showing the circumferential ridge, very strongly pronounced (oblique and side-view). (2) Diskoid (length 6.5 cm, weight 145.02 g) showing a wavy median ridge, not very pronounced (side-view). (3) Doughnut-shaped form, plan view showing central depression, present on both sides; side-view showing circumferential ridge. (4) Ovoid form, plan view showing very small central dimple; side-view showing circumferential ridge. (5) Dumbell form (length 7.5 cm, width at waist 3 cm, weight 123.4 g), plan view; side-view showing very faint circumferential ridge.

the target rock stratigraphy, and moldavite tektites have been interpreted as high-speed ejecta from the uppermost layers of the target rocks, a view supported by later chemical and isotopic studies (Koeberl 1994). The major and trace elements of the tektites can be explained if they are derived from the Upper Freshwater Molasse unit, but not any deeper stratigraphic unit. Sm:Nd and Rb:Sr isotope ratios support this.

Bosumtwi Crater, Ghana
This crater (Figure 1a), with a centre at 6° 30′ N, 1° 25′ W, has a diameter of 10.5 km and was excavated in metamorphic rocks (lower greenschist facies metasediments) dated at about 2000 million years (Schnetzler *et al.* 1966). The target rocks also include granitic rocks (Jones 1985). There is a good chemical correlation between the target rocks, impactites at the crater and the Ivory Coast tektites. Glass *et al.* (1991) showed that the geographic concentration in deep-sea cores of microtektites off the West African coast is consistent with a source in the crater. The stratigraphic position of the microtektites (about 0.97 million years on geomagnetic evidence) is consistent with the age of the tektites (1.04 $\pm$0.01 million years) and the crater (about 1 million years) (Jones *et al.* 1981).

Chesapeake Structure
A complex 'peak–ring' structure has recently been discovered beneath Chesapeake Bay, its bordering peninsulas and the adjacent inner continental shelf, centred on 37° 16.5′ N, 76° 0.7′ W, and measuring 90 km in diameter (Anon 1995; Poag 1996; Koeberl *et al.* 1996) (Figure 1b). The Exmore Breccia filling the 'crater' displays shocked quartz, plagioclase and orthoclase. The compositions of some of the high-silica rocks match well with those of North American tektites for most non-volatile elements. The age of the structure is about 35 million years, consistent with the formation age of North American tektites and the formation in Barbados in which tektite fragments have been recovered. It would appear likely that this is the source of the North American strewn field.

Australasian strewn field
No entirely convincing source has been found. The search has been constrained by the idea that large, irregular, blocky, layered and

poorly homogenized Muong Nong-type tektites must be near the source, and that this gave rise, because of their spread, to the multiple source hypothesis of Wasson (1991). This has been rejected on the basis of field examination by Fiske *et al.* (1996) near the town of Huai Sai in north-east Thailand, and the chemical consistency of tektites surely must also rule it out. Also, in two strewn fields layered Muong Nong types occur over an extent of 2000 km within the strewn field, and it seems that, whereas the case for them being less perfectly homogenized forms has been established, they do not have to be found immediate to the source. The distribution of large tektites in the Australasian strewn field is also not straightforward. The large Muong Nong forms of Indochina are in a general sense proximal – irregular, blocky and layered forms are not reported from Australia, which must surely be at the distal end of the strewn field – but tektite distribution is very irregular, and indeed there may be zones characterized by certain splash-form shapes or chemical types, both radially and concentrically (Stauffer 1978). Although many Australian tektites are small, more than seventy over 100 g have been recorded (Figure 7); the largest weighs 437.5 g (Cleverly 1994) and must have been projected thousands of kilometres from the source. Enormous numbers of tektites have been found at Vietnamese sites, hundreds of thousands at Dalat alone (Schnetzler 1992), and this, as well as the size factor, has led to the search for a source in Indochina. Sources in Laos, Cambodia and Vietnam all have been suggested (for example Stauffer 1978; Hartung & Rivolo 1988; Burns & Glass 1989; Hartung 1990; Hartung & Koeberl 1994; Koeberl 1992, 1995; Schnetzler 1992; Schnetzler & McHone 1996) and also offshore from Vietnam in the 100-km diameter Qui Nhon gravity anomaly (Schnetzler *et al.* 1988). Studies of the distribution of microtektites in deep-sea cores have favoured a source in Indochina (Glass & Wu 1993; Glass & Pizzuto 1994; Glass *et al.* 1996). However, the most favoured source, the large Tonle Sap linear lake depression in Cambodia, has so far yielded no convincing evidence, and the source structure of the immense Australasian strewn field (Figure 1a), only 0.7–0.8 million years old, which should surely be at least as large as the Chesapeake Structure (90 km), would surely be easy to recognize if on land, for it could not have suffered much degradation by erosion, though it could have been obscured by later volcanic formations or loess.

Tektites, mainly of Muong Nong type, are found right up

Figure 7. Two very large australites. Above: a rounded core from Lake Grace, Western Australia (Western Australian Museum Collections). This weighs 168.4 g. It has lost all traces of ablation, though the posterior and anterior surfaces can still be recognized. The anterior surface is marked by numerous U-grooves. Below: a large dumbell-shaped australite from Hyden, Western Australia (described by Cleverly 1990). This weighs 130.98 g and is 11.33 cm long, being the longest australite specimen recorded. It is marked by numerous sinuous U-grooves. This is a side-view with the posterior surface towards the top. These are both ablated tektites. (Photographs by W. H. Cleverly.)

to Vietnam's northern border with the Yunnan and Guangxi Provinces of China (Schnetzler 1992; Schnetzler & McHone 1996), and the source might well be in China. Tektite studies do not appear to have been pursued to any extent in these provinces, though they have been vigorously pursued in Guangdong and Hainan Island. It is interesting to note that all the supposed sources of the three other strewn fields are at least 300 km from the nearest tektite finds, so perhaps the abundance of tektites in Indochina might be taken as an indicator that the source is elsewhere.

The unanswered questions

Exhaustive researches on tektites, backed up by studies of tektite fragments and microtektites in deep-sea drill cores, have delineated the strewn fields in reasonable detail and have indicated a terrestrial, near-surface sedimentary rock source, ruling out an extraterrestrial source. The source of the Australasian strewn field discussed above is unquestionably the most important (Taylor 1996). Yet unanswered questions remain.

1) Why are all the source locations in the Northern Hemisphere? (This may or may not be significant.)
2) The less complete homogenization of Muong Nong-type layered tektites has been established, but the consistent binary component nature is surprising, and the possibility that liquid immiscibility is involved (Zolensky & Koeberl 1991) calls for further investigation.
3) The large size and blocky, irregular nature of some Muong Nong-type tektites in Indochina seems to be due to breaking off from larger parent masses, sintering into large masses on the ground being excluded. The Muong Nong type is defined by its layered *and* blocky irregular character. Unfortunately, there does not seem to have been any systematic study of the distribution of large, irregular blocky Muong Nong forms in Indochina, Thailand, China and the Philippines (a study such as that by Cleverly (1994) would be invaluable). Is it possible that even such large masses (12 kg and more) have been projected high into the atmosphere and even re-entered? It seems that this possibility should not be rejected out of hand, in view of the close association of large blocky Muong Nong forms with intermediate layered types showing splash forms (Izokh & An 1988; Schnetzler 1992). An alternative possibility is put forward by Koeberl (1994), who states that 'some tektites solidify in near-vacuum and re-enter the atmosphere. During the re-entry they melt again and form ablation-shaped tektites. Larger tektites from the lower part of the target stratigraphy are only distributed closer to the source crater.'
4) There is a need to investigate systematically, in further detail,

the residual traces of ablation (equatorial scars and ridges, ablation surfaces) on tektites from South-east Asia and China and from the other three strewn fields, and in particular on large individual tektites, including the blocky Muong Nong types. The collections from Indochina, including the immense Dalat and Da Nang collections from Vietnam, should be included, as should those from the Philippines and the lesser known collections from China. Such studies should be coupled with further experiments, computer simulations and ballistic studies of ablation, with particular emphasis on the ablation behaviour of large tektites and consideration of the limitations on the projection of large tektite bodies. The aim here is to resolve the question of whether some large tektites never suffered ablation, or whether all tektites are bodies or fragments of bodies that suffered ablation. There is another question, related to ablation, one that an extreme sceptic might pose: 'Is there any other ablationary process that could have moulded ablated tektites other than projection up from the Earth, solidifying in near-vacuum and atmospheric re-entry?' The answer to this would seem to be a firm 'No!', but tektites are extraordinary objects and the question may yet require an extraordinary answer.

5) The question of why, among hundreds of thousands of tektites collected, we never find undigested source rock also needs considering. Perhaps not enough of the less well-homogenized Muong Nong types have been sliced, but the reason may rather be a fundamental one (Wasson 1996).
6) Why are there no tektite developments associated with more than four of the 150 or so cryptoexplosion structures and craters widely attributed to meteorite impacts (Grieve 1991; Koeberl 1994; Taylor 1996)? Terrestrial degradation or obscuration with age is not an entirely satisfactory answer, for there are many other such structures within the age range of tektites. Do only target sediments of a certain type yield tektites? Is there some constraint in the dynamics and angle of impact, or the nature of the impacting body (Wasson 1996)? Drawing an analogy with obsidian, which should decompose more easily than tektites – which are extremely dry glasses – the expectation would be for tektite layers older than 35 million years to be preserved in sedimentary sequences, but none are known of tektites. Perhaps a more rigorous search is needed.

7) Why do all strewn fields have a sector distribution away from the source (though the three suggested source structures are circular)? Oblique ejection rather like the divot from a golf stroke has been suggested, but there may be another explanation.
8) Do we have to limit our arguments to the restricted range of familiar meteorite types (implying asteroids)? Hamilton (1993) has suggested that meteorites which strike the Earth sample only a limited part of the asteroid belts of the Solar System, and cometary impacts are gaining favour (for example Shoemaker 1997). There may be an analogy here with the Tunguska event in Siberia in 1908, which left no trace of the impacting object except some cosmic dust particles, and may well have been cometary (Wasson 1996).
9) There remains the possibility, suggested by several authors (for example Currie 1969, 1972; Officer & Carter 1991; Officer 1992), that not all cryptoexplosion structures associated with shock effects and high-pressure polymorphs of minerals are of extraterrestrial impact explosion origin. There are indeed some remarkable clusterings of such structures, and some remarkable absences over large areas (for example, in China), but this may be a collection bias. Tektites are certainly *not* products of terrestrial volcanism in the guise familiar to geologists, and, though there remains the remote possibility that there are cryptic shock processes of endogenous origin that have been confused with impact explosion processes, the whole character of tektites would seem to rule out anything but an origin in projection of *terrestrial target material* from sites of impact of large *extraterrestrial* bodies.

While these questions remain, however, tektites will continue to constitute a scientific enigma.

References

Anon 1995. 'Making an impact under Chesapeake', *Geotimes*, January 1995, 7–8.
Barnes, V. E. 1940. 'North American tektites – contributions to geology', Univ. of Texas Publ. No. 3945, 477–656.
Barnes, V. E. 1969. 'Progress of tektite studies in China', *EOS*, **50**, 704–708.
Baker, G. 1963. 'Form and sculpture of tektites', in J. A. O'Keefe (ed.), *Tektites*, Univ. of Chicago Press, 1–24.

Burns, C. A. & Glass, B. P. 1989. 'Source region for the Australasian strewn field', *Meteoritics*, **24**, 257.
Chapman, D. R. & Larson, H. K. 1963. 'Lunar origin of tektites', NASA Tech. Note D1556, Washington, DC, 66pp.
Cleverly, W. H. 1988. 'Australites from Mt. Remarkable Station and adjacent parts of Yerilla Station, Western Australia', *Records of the Western Australian Museum*, **14**, 225–235.
Cleverly, W. H. 1990. 'A large dumbell-shaped australite from north of Hyden, Western Australia', *Records of the Western Australian Museum*, **14**, 661–663.
Cleverly, W. H. 1994. 'The heaviest known australites (Australian tektites)', *Western Australian School of Mines Magazine*, 77.
Currie, K. L. 1969. 'Geological notes on the Carswell circular structure, Saskatchewan (14K)', Canada Geological Survey Paper 67–32, 60pp.
Currie, K. L. 1972. 'Geology and petrology of the Manicouagan cryptoexplosion structure, Quebec, Canada', Canada Geological Survey Bull. 198, 153pp.
Darwin, C. 1844. *Geological Observations on the Volcanic Islands Visited During the Voyage of H.M.S. 'Beagle'*, Smith, Elder & Co., London.
Fiske, P. S., Prinya Putthapiban & Wasson, J. T. 1996. 'Excavation and analysis of layered tektites from Northeast Thailand: Results of 1994 field expedition', *Meteoritics & Planetary Science*, **31**, 36–41.
Glass, B. P. 1989. 'North American tektite debris and impact ejecta from DSDP Site 612', *Meteoritics*, **24**, 209–218.
Glass, B. P. 1993. 'Geographic variations in abundance of Australasian microtektites: Implications concerning location and size of the source crater', *Meteoritics*, **28**, 354.
Glass, B. P., Chapman, D. R. & Shyam Prasad, M. 1996. 'Ablated tektite from the Central Indian Ocean', *Meteoritics & Planetary Science*, **31**, 365–369.
Glass, B. P., Kent, D. V., Schneider, D. A. & Tauxe, L. 1991. 'Ivory Coast microtektite strewn field: Description and relation to the Jaramillo geomagnetic event', *Planetary Science Letters*, **107**, 182–196.
Glass, B. P. & Pizzuto, J. E. 1994. 'Geographic variation in Australasian microtektite concentrations: Implications concerning the location and size of the source crater', *Journal of Geophysical Research*, **99**, 19075–19081.

Glass, B. P. & Wu, J. 1993. 'Coesite and shocked quartz discovered in the Australasian and North American microtektite layers', *Geology*, **21**, 435–438.

Grieve, R. A. F. 1991. 'Terrestrial impact: The record in the rocks', *Meteoritics*, **26**, 175–194.

Hamilton, W. B. 1993. 'Evolution of the Archaean mantle and crust', in *The Geology of North America, Vol. C-2, Precambrian: Conterminous U.S.*, Geological Society of America, 597–614, 630–636.

Hartung, J. B. 1990. 'Australasian source crater: Tonle Sap, Cambodia', *Meteoritics*, **25**, 369–370.

Hartung, J. B. & Koeberl, C. 1994. 'In search of the Australasian source crater: The Tonle Sap hypothesis', *Meteoritics* **29**, 411–416.

Hartung, J. B. & Rivolo, A. R. 1979. 'A possible source in Cambodia for Australasian tektites', *Meteoritics*, **14**, 153–160.

Horz, F. 1982. 'Ejecta of the Ries crater', in L. T. Silver & P. H. Schultz (eds), *Geological Implications of Impacts of Large Asteroids and Comets on Earth*, Boulder, Col., Geological Society of America Special Paper 190, 39–45.

Izokh, E. P. & An, L. D. 1983. 'Tektites of Vietnam: Tektites delivered by a comet – a hypothesis', *Meteoritika*, **42**, 158–160.

Jones, W. B. 1985. 'Chemical analyses of Bosumtwi crater target rocks compared with Ivory Coast tektites', *Geochimica et Cosmochimica Acta*, **49**, 2569–2576.

Jones, W. B., Bacon, M. & Hastings, D. A. 1981. 'The Lake Bosumtwi impact crater, Ghana', *Geological Society of America Bulletin*, **92**, 342–349.

Koeberl, C. 1992. 'Geochemistry and origin of Muong Nong-type tektites', *Geochimica et Cosmochimica Acta*, **56**, 1033–1064.

Koeberl, C. 1994. 'Tektite origin by hypervelocity asteroidal or cometary impact: Target rocks, source craters and mechanisms', Geological Society of America Special Paper 293, 133–153.

Koeberl, C. & Beran, A. 1988. 'Water content of tektites and impact glasses and related chemical studies', *Proc Houston Lunar & Planetary Science, Houston Lunar & Planetary Science Institute*, **18**, 403–408.

Koeberl, C. & Glass, B. P. 1988. 'Chemical composition of North American microtektites and tektite fragments from Barbados and DSDP Site 612 on the continental slope of New Jersey', *Earth and Planetary Science Letters*, **87**, 286–292.

Koeberl, C., Poag, C. W., Reimold, W. U. & Brandt, D. 1996. 'Impact origin of the Chesapeake Bay structure and the source of the North American tektites', *Science*, **271**, 1263–1266.

Lacroix, A. 1932. 'Les tectites de l'Indochinie', *Archives de la Musée d'Histoire Naturelle, Paris*, **6** (Ser. 8), 193–196.

Lacroix, A. 1934. 'Sur la découverte de tectites à la Côte d'Ivoire', *Comptes Rendus, Académie des Sciences, Paris*, **199**, 1539–1542.

McCall, G. J. H. 1973. *Meteorites and their origins*, David & Charles, Newton Abbot, 352pp.

McCall, G. J. H. & Cleverly, W. H. 1969. 'The Nallah meteorite, Western Australia – A small oriented common chondrite showing flanged button australite form simulation', *Mineralogical Magazine*, **37**, 286–287.

McNamara, K. & Bevan, A. W. R. 1991. *Tektites*, 2nd edn, Western Australian Museum, Perth, 28pp.

Officer, C. B. 1992. 'The relevance of iridium and microscopic dynamic deformation features towards understanding the Cretaceous–Tertiary transition', *Terra Nova*, **4**, 394–404.

Officer, C. B. & Carter, N. L. 1991. 'A review of the structure, petrology and dynamic deformation charcteristics of some enigmatic terrestrial structures', *Earth Science Reviews*, **30**, 1–49.

Poag, C. W. 1996. 'Structural outer rim of Chesapeake Bay impact crater: Seismic and borehole evidence', *Meteoritics & Planetary Science*, **31**, 218–226.

Sanfilippo, A., Reidel, W. R., Glass, B. P. & Kyte, F. T. 1985. 'Late Eocene microtektites and radiolarian extinctions in Barbados', *Nature*, **314**, 613–615.

Saunders, J. B., Bernouilli, D., Muller-Merz, F., Oberhansli, H., Perch-Nielsen, K., Riedel, W. R., Sanfilippo, A. & Torini, R. 1984. 'Stratification of the late Middle Eocene – early Oligocene in the Bath Cliff section, Barbados, West Indies', *Micropalaeontology*, **30**, 390–425.

Schnetzler, C. C. 1971. 'The lunar origin of tektites – R.I.P.', *Meteoritics*, **5**, 221–222.

Schnetzler, C. C. 1992. 'Mechanism of Muong Nong-type tektite formation and speculation on the source of the Australasian tektites', *Meteoritics*, **27**, 154–165.

Schnetzler, C. C. & McHone, J. F. 1996. 'Source of Australian tektites: Investigating possible impact sites in Laos', *Meteoritics & Planetary Science*, **31**, 73–76.

Schnetzler, C. C. & Pinson, W. H. 1963. 'The chemical composition of tektites', in J. A. O'Keefe (ed.), *Tektites*, Univ. of Chicago Press, 95–129.

Schnetzler, C. C., Pinson, W. H. & Hurley, H. 1966. 'Rubidium–strontium age of the Bosumtwi crater area, Ghana, compared with the Ivory Coast tektites', *Science*, **151**, 818–819.

Schnetzler, C. C., Waller, L. S. & Marsh, J. G. 1988. 'Source of the Australasian tektite strewn field – A possible offshore impact site', *Geophysics Research Letters*, **15**, 357–360.

Shoemaker, E. M. 1997. 'Long term variations in the impact cratering rate on Earth', *Geoscientist*, **7**, 29.

Stauffer, P. H. 1978. 'Anatomy of the Australasian strewn field and the probable site of its source crater', Proceedings of the 3rd Regional Conference on Geology and Mineral resources of SE Asia, Asia Institute of Technology, Bangkok, 285–289.

Taylor, S. R. 1973. 'Tektites – A post-Apollo view', *Earth Science Reviews*, **9**, 101–123.

Taylor, S. R. 1996. 'Tektites – Some unresolved problems', *Meteoritics & Planetary Science*, **31**, 4–5.

Taylor, S. R. & Koeberl, C. 1994. 'The origin of tektites: Comment on a paper by J. A. O'Keefe', *Meteoritics*, **29**, 739–744.

Wasson, J. T. 1995. 'Layered tektites – A multiple impact origin for the Australasian tektites', *Earth & Planetary Science Letters*, **102**, 95–109.

Wasson, J. T. 1995. 'The disintegration of the comet Shoemaker–Levy 9 and the Tunguska object and the origin of Australasian tektites', *Lunar & Planetary Science*, **26**, 1469–1470.

Wenzhli Lin, Ziyuan Ouyang & Shijie Wang 1995. 'Cosmochemistry in China', *Episodes*, **18**, 95–97.

Xu Dao-Yi, Yan Zheng, Sun Yi-Yin, He Jin-Wen, Zhang Qin-Wen & Chai Zhi-Fang 1989. 'Tektites in the Early Quaternary of China', in *Astrogeological Events in China*, Geological Publishing House, Beijing; Van Nostrand Reinhold, NY; and Scottish Academic Press, Edinburgh, 156–170.

Zolensky, M. E. & Koeberl, C. 1991. 'Why are blue zhamanshinites blue? Liquid immiscibility in an impact melt', *Geochimica et Cosmochimica Acta*, **55**, 1483–1486.

Postscript: the author's long-time friend and colleague, W. H. Cleverly, was taken ill while collecting tektites in the field in early 1997 and died shortly afterwards. This short review article is dedicated to his memory.

Dust Rings around Normal Stars

HELEN J. WALKER

One of the most exciting discoveries made by the Infrared Astronomical Satellite (IRAS) was the ring of dust around a normal, nearby star, Vega.[1] When Vega was observed by IRAS, it was brighter in the infrared than expected due to the extra emission from the cool dust ring around the star. The temperature of the dust was around −200°C. Although Vega is twice as hot as the Sun, like the Sun it does not blow off large amounts of material, so H. H. Aumann (Jet Propulsion Laboratory, Pasadena) and F. C. Gillett (Kitt Peak National Observatory, Tucson) concluded that the dust ring must have been around Vega since the star formed. They also deduced that the dust grains in the ring were larger than the dust grains normally found in interstellar space. However, Vega is not the type of star that should have dust around it.

Vega is on the main sequence of stars (as is the Sun). The main sequence can be used to relate the temperature of a star to its gravity; a red giant like Betelgeuse, well above the main sequence on the Hertzsprung–Russell diagram, has a lower gravity than the main sequence indicates according to its temperature. New stars are formed when a cloud of gas and dust collapses. They have a lot of dust around them, and a large ring of gas and dust is usually formed. As the star collapses, and the nuclear burning starts (converting hydrogen into helium), the star is both gaining material from the ring and losing it to the surrounding cloud. The star loses the dust ring before it arrives on the main sequence, as well as the cocoon from which it was born, so we can easily see the star. Old stars create dust and blow the material out from their atmospheres when they have left the main sequence, and so a cloud of dust and gas forms around the star. Eventually a substantial proportion of the original mass of the star is in the cloud around the star. Planetary nebulæ and supernovæ can occur when the conditions are right, and then very large amounts of material are 'suddenly' blown away from the star. Consequently, a star on the main sequence should not still have a ring of dust around it – the dust should gradually drift away, and no new dust is fed into the ring from the star.

After the discovery of Vega's dust ring, the IRAS data for other main sequence stars were examined for more examples, and three stars were soon identified as showing the same type of ring: Alpha Piscis Austrini (Fomalhaut), Beta Pictoris and Epsilon Eridani. These four stars are regarded as the prototypes of a class known as Vega-like stars. They are all relatively bright, nearby stars. Since the initial discovery of the prototypes, extensive searches of the IRAS data using different techniques, by several groups, have produced a list of about 30 or 40 candidates with cool dust around them. Of the four prototypes, only for Beta Pictoris has the ring been observed in visible light. The ring is seen edge-on, made visible by light scattered from the star itself. The ring around Vega, by contrast, is deduced to be face-on, since there is no evidence that light from Vega itself has been affected by passage through dust.

Beta Pictoris is unique in another way, in that it is the only prototype for which lines from ionized gas have been detected. A. M. Lagrange-Henri (Université J. Fourier, Grenoble), A. Vidal-Madjar (CNRS, Paris) and co-workers observed Beta Pictoris from the ground and via the International Ultraviolet Explorer (IUE) satellite. They found that lines from ionized calcium, magnesium and other materials were present all the time, but sometimes there were additional redshifted lines. The Hubble Space Telescope (HST) has also observed these lines of ionized gas from the environment around Beta Pictoris, including some redshifted lines. These lines may be indicating that material from the ring occasionally falls onto the star. The models suggest that we are seeing kilometre-sized bodies evaporating as they approach the star, and that the frequency of these comets falling on to the star could be higher than a hundred per year. In this fashion, the gas ring around the star is continually replenished, despite gas being blown away by the star's stellar wind (like the solar wind).

The theories developed to explain the observations of the four prototypes lead us to a view of these dust rings which reminds us of our own Solar System, with its Kuiper Belt and Oort Clouds beyond the planets. The dust rings are cool (all around −200°C). They have a gap in the centre nearest the star, where there is no dust, around 50 AU across. The ring extends out to around 500 AU in diameter for Vega, and 1500 AU for Beta Pictoris, according to the theories. It is very difficult to calculate how much dust is present in the rings. As a dust 'grain' gets larger it emits less efficiently, so that we could not see an Earth-sized planet at these distances with IRAS, but if

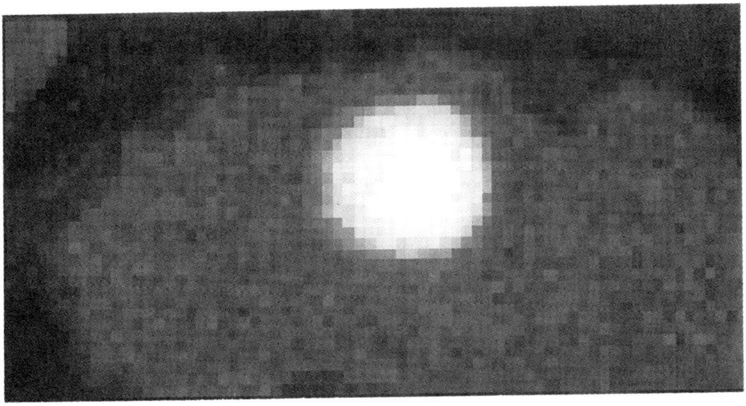

Figure 1. Picture of Vega observed by the ISO photometer, at 60 microns. The image is larger than that expected for the star alone (about 100 arcsec across).

that same material is spread out as small dust grains, it is easily detected. In the case of Vega, if the biggest dust grain were 2 mm in diameter, the amount of dust in the ring would be one-hundredth the mass of the Earth, and if the biggest dust 'grain' were the size of an asteroid (say 200 km), there could be a hundred times the mass of the Earth in the dust ring. Beta Pictoris has a narrower range of estimates for the amount of material in the ring, ranging from 1 to 5 Earth masses. If we compare the mass of the Earth with (just) the mass of Jupiter, it tells us that there is a lot less dust around these Vega-like stars than is present in our own Solar System.

Since the dust is so cool, it emits energy mainly in the infrared, and IRAS could find it easily. From the ground, there are very few windows in the atmosphere through which to observe the dust rings, so follow-up work since IRAS has focused on the region around 10 μm and at millimetre wavelengths, although neither is as appropriate for detecting the energy as is the infrared range observable from space by IRAS or its successor ISO (the Infrared Space Observatory). The launch of ISO by the European Space Agency, on November 17, 1995, has given us a new opportunity to study these dust rings. At the heart of ISO is a 0.6-m telescope and many very sensitive infrared detectors. The whole telescope is cooled by liquid helium, and this will enable us to make observations for about two years. There are two spectrometers on board, covering the range 2.5–180 μm, and these will give us a totally new view of the dust (and any gas) and the chemical processes at work in the dust rings by showing us which atoms and molecules are present, as well how many there are and their temperatures. In addition, the photometer on ISO measures the energy (brightness) of the dust at various wavelengths between 2.5 and 200 μm, and can make small maps to show the distribution of the cool dust seen by IRAS. There is also a small camera on ISO which can make higher-resolution maps of the dust rings at shorter wavelengths, from 2.5 to 17 μm.

The ISO photometer has been used to observe the region around Vega, and the map obtained gives an image which is bigger than that predicted for the star alone, so we are finally seeing the ring around Vega. I lead a team of scientists, associated with the photometer, in a project to study some of the dust rings found with the IRAS satellite. We still have to remove the natural blurring due to the passage of the infrared radiation through the telescope and instrument (termed the instrumental profile), but we think the ring could be as large as 700 AU across (larger than suggested from theories based on the IRAS data). We used a different technique on Fomalhaut, and we think the ring there could be as much as 800 AU across (closer to the figure those theories suggested). These numbers are rather similar to that for the (inner) Oort cloud believed to surround our own Solar System. Vega is about 26 light years away and Fomalhaut 24 light years, but there are other candidates farther away. These show similar characteristics, and again one or two rings may have been imaged, one from a star about 50 light years away.

The millimetre wavelengths have proved very interesting for looking at the coolest dust around the candidate stars, to see how large the systems can become. The prototype Vega-like stars are very faint at millimetre wavelengths, but Rolf Chini's group (Max Planck Institut für Radioastronomie, Bonn) used the 30-metre IRAM telescope in Spain to detect Fomalhaut, and found it to be extended, so they were seeing the dust ring. There are several groups involved in this work, and their results do not agree. However, when the sizes of the different telescopes are considered, and the different instrumentation, the disagreements often mean that one telescope has not seen the whole dust ring, and another telescope has, so the disagreements have proved very useful.

We can start to investigate the composition of the dust at a wavelength of 10 μm, using the large infrared telescopes in Hawaii, UKIRT or the NASA IRTF, as well as ISO. Ground-based observations of Beta Pictoris showed that the dust was made of silicate

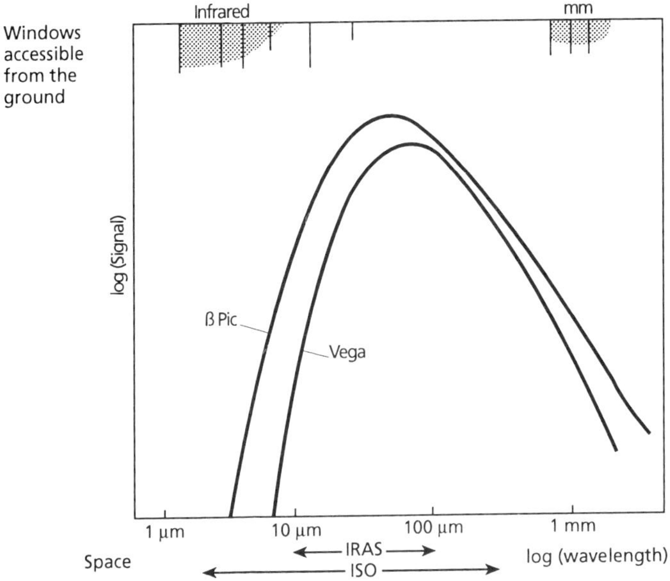

Figure 2. The energy emitted by the dust rings around Vega and β Pic are shown. Also shown is the wavelength region covered by IRAS and by ISO, compared to the windows accessible from the ground in the infrared and millimetre ranges.

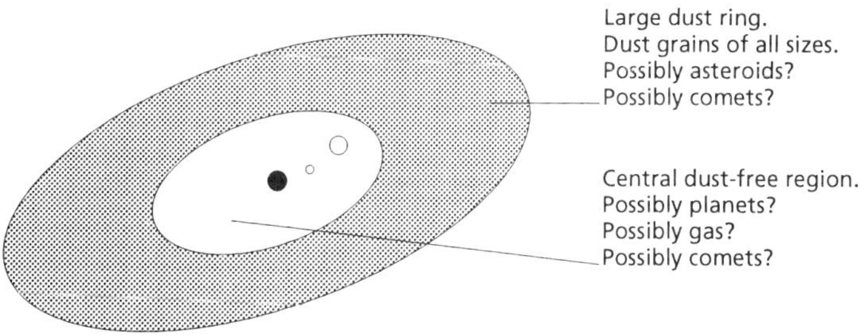

Figure 3. An artistic impression of a Vega-like system, showing the various components thought to be there.

material, with a range of particle sizes. However, the spectra show additional features which suggest that the silicate material could be similar to that detected in Comets Halley and Bradfield. The dust emission from Vega and Fomalhaut is too faint at 10 μm for any features to be observed. Other candidate stars have also been shown to have silicate material around them. One shows a spectral feature from silicate dust which is very similar to that found for Beta Pictoris, whilst another candidate has a silicate dust feature with a different shape, much more like that caused by the dust around young T Tauri stars. Two candidate stars have been found by Mike Barlow and Roger Sylvester (University College, London) and colleagues, using UKIRT, to show the signature of complex organic molecules (polycyclic aromatic hydrocarbons, PAHs). The small spectrometer in the ISO photometer has confirmed the signature of these complex organic molecules from these two stars, plus a third.

When the dust ring around Vega was discovered, it was thought to represent a failed attempt at producing planets. The theories suggested that the dust grains around Vega were larger than those in interstellar space, implying that the process of dust grains sticking together and growing had started, but that there was not enough dust in the ring for the process to get very far, and that no planetary bodies had formed. As mentioned earlier, the total amount of dust around Vega is comparable to the mass of the Earth, which is a small fraction of the total mass of the Solar System. Many theories of the Solar System's formation have difficulties explaining how planets form around a young star before the usual large dust ring is blown away by the star itself. Recently a class of young stars has been discovered (called naked T Tauri stars), which suggests that some stars may never have developed a dust ring, or that they got rid of it very early in their lifetime. The Vega-like stars show that some stars manage to retain a ring of dust for a very long time. Whilst some stars, like Vega, have a very tenuous ring, others may have more massive rings, with enough dust to make a solar system with planets. These stars may be good places to look for Earth-sized planets in years to come, when larger telescopes are flown on board satellites. The detection of the complex organic molecules should not be interpreted as a sign that these planets, if they exist, will contain life, but they have shown us that the complex organic molecules are very tough, surviving in a wide range of conditions in interstellar space, including proximity to a star which could heat them severely.

ISO will be used by many groups to observe Vega-like stars during its two-year mission. We want to find out more about the prototypes, confirm whether the candidates proposed really are true Vega-like stars, or something closer to the young T Tauri stars, and we want to look for new candidates. There will be time, after ISO runs out of its helium coolant, to apply the theories to the new data, and to develop more detailed theories. We cannot use ISO to observe Beta Pictoris at any time. It can look at only a limited part of the sky at any time, since it must never look at the Sun, and must avoid the Earth and Moon, so the part of the sky near Orion (where Beta Pictoris is), has only been accessible from July 1997 – a suitable finale to the project.

Note

1. See the 9th Sky at Night book for more information on the discovery and the IRAS mission.

On the Brink of Black Holes – The Story of Neutron Stars

CHRIS KITCHIN

Imagine a cold winter's night and sitting, toasting your toes, before a roaring log fire. The flickering flames leap up the chimney, dragging the fumes of the combustion with them, while the heat radiates out into the room. But why do the flames rise up the chimney? The answer is basic physics: when a gas is heated it expands, or if it is in a confined space, then its internal pressure rises.

This process is so much the natural order of things that we accept the consequences without any thought. But does it always have to happen? The answer is no: there are materials in which the pressure remains constant when they are heated or cooled. If we lived in a world composed of such material, then it would be drastically different from anything we presently experience. Any process depending on combustion or convection would no longer work. There would be no fires, no internal combustion engines, no weather processes, and we ourselves would have temperatures rising towards 100°C as our bodies had to rely just on conduction and radiation to keep us cool.

The effect on the Sun at first sight would seem to be that it becomes larger and possibly brighter because radiation and not convection would now have to transport the energy through its outer layers. In fact the effect on the Sun would be much more fundamental. The Sun is composed of a hot plasma, with the temperature at its centre about 15 million °C. That enormous temperature leads to an equally high pressure, which enables the material in the core of the Sun to support the huge weight of its outer layers. But if the pressure of a plasma did not depend upon the temperature, so that at 15 million °C the pressure were the same as at the surface of the Sun, where the temperature is a mere 6000°C, then the pressure at the centre would be insufficient to support the outer layers, and the Sun would collapse. If nothing else intervened, then in about a quarter of an hour the Sun would have shrunk to half its present radius, and five minutes or so after that it would have become a black hole. We may thus be grateful that we are associated with material in which pressure does depend on temperature.

Material in which presure does not depend on temperature comes in two forms, and its properties lead to two of the most intriguing objects to be found in the sky: white dwarfs and neutron stars. The material forming white dwarfs is called electron-degenerate matter, while that forming neutron stars is called baryon-degenerate[1] matter. The reason why the pressure in such material does not depend on its temperature has to do with some intricate and obscure aspects of quantum mechanics. In electron-degenerate material, essentially all available niches for the electrons are already occupied. When the material is heated, the normal effect would be for the electrons to move faster. But in electron-degenerate material the niches representing those faster velocities are already full, so the electrons have to remain at their old velocities. Since the electron pressure depends on the electrons' velocities, it therefore does not change when electron-degenerate matter is heated. In baryon-degenerate matter, it is the neutrons and protons whose velocities are fixed, and their pressure similarly remains constant.

Now, the total pressure inside an object is the sum of pressures arising from various sources, such as gas pressure, electron pressure, radiation pressure and the pressure from protons, neutrons and other particles. So imagine the Sun towards the end of its life, some 5 billion years from now. The nuclear reactions will have halted, and it will be collapsing slowly as it radiates away its stored energy. The major source of pressure supporting the overlying layers will be that due to degenerate electrons. But this pressure, as we have seen, does not depend on the temperature. In fact it depends only on the density of the material. The collapsing Sun will therefore stabilize only when it has shrunk to about the present size of the Earth, and its average density has risen to 2 million times that of water (or about 60 tonnes compressed into a matchbox – Figure 1). If we add more material to the Sun at this stage, then the weight of that extra material has to be supported by an increase in the internal pressure, and that can only come from a further increase in density. The result is that the white dwarf, for that is what the Sun will then have become, shrinks as it becomes more massive. At a mass about 1.4 times the mass of the Sun, it will theoretically have decreased to zero size. This mass therefore represents an upper

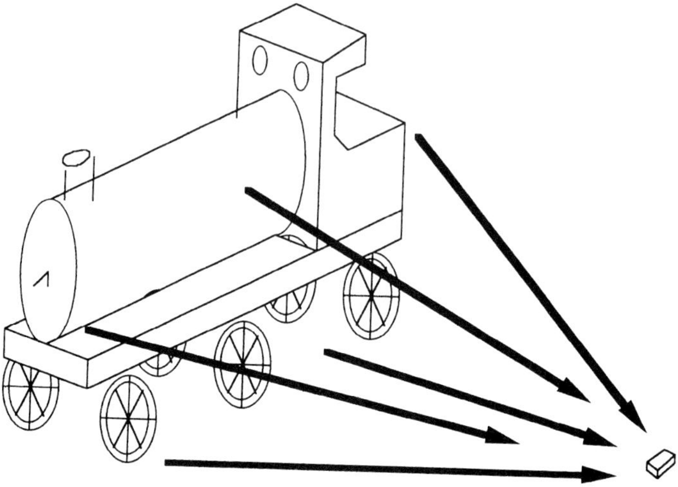

Figure 1. 60 tonnes (e.g. a steam engine) compressed into a matchbox.

limit to the possible masses for white dwarfs, and is termed the Chandrasekhar limit.

Thus stars which are significantly more massive than the Sun, if they retain that mass to the end of their lives, cannot end as white dwarfs. Instead they must continue to contract as they radiate away their energy. If they are massive enough, they will end as black holes. However, between the Chandrasekhar limit and some 2 to 3 times the mass of the Sun, the collapse can be halted by the rising pressure of baryon-degenerate material. The pressure again depends only on density, and becomes sufficient to stabilize the collapsing star when the density is close to that of an atomic nucleus (2×10^{17} kg/m^3, or about 5 billion tonnes in a match box – Figure 2).

The size of a neutron star with a mass twice that of the Sun is about 20 km. A static black hole[2] of the same mass would be about 10 km across. So neutron stars are not that much larger than their corresponding black holes, and the escape velocity from the surface of the neutron star would be over three-quarters of the speed of light.

To anyone versed in modern physics, there is an immediate problem with the idea of a star composed of neutrons: the neutron is an unstable particle with a half-life of about 11 minutes. How, then,

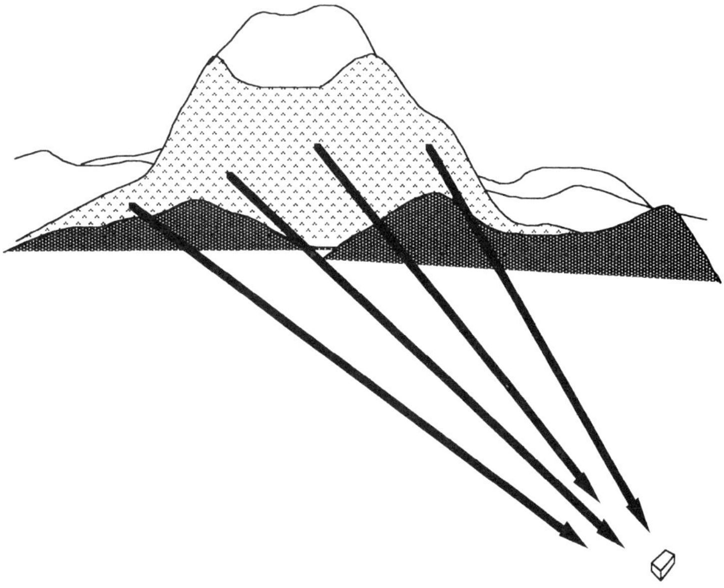

Figure 2. Five billion tonnes (e.g. a large mountain) in a matchbox.

can neutron stars exist for any length of time without many of the neutrons changing into protons and electrons? The answer is that there is no room for them to decay back to protons and electrons. The neutrons are embedded in a sea of degenerate electrons, so that (as when an electron-degenerate gas is heated) there are no niches left for the electrons produced by neutron decay to fill. They therefore have to remain locked inside the neutrons.

The existence of neutron stars was predicted in 1934 by Walter Baade and Fritz Zwicky, only two years after the neutron itself was discovered. However, the first evidence that neutron stars were more than just a theoretician's plaything did not come for over three decades. In 1967 Jocelyn Bell found a signal recurring at intervals of 23 hours 56 minutes in the output she was monitoring from a new radio telescope at Cambridge. This time interval corresponds to a sidereal day, indicating that the source of the signal was almost certainly an astronomical object, and not terrestrial interference. This was important because the signal itself looked highly artificial – a series of radio pulses at very precise and regular time intervals

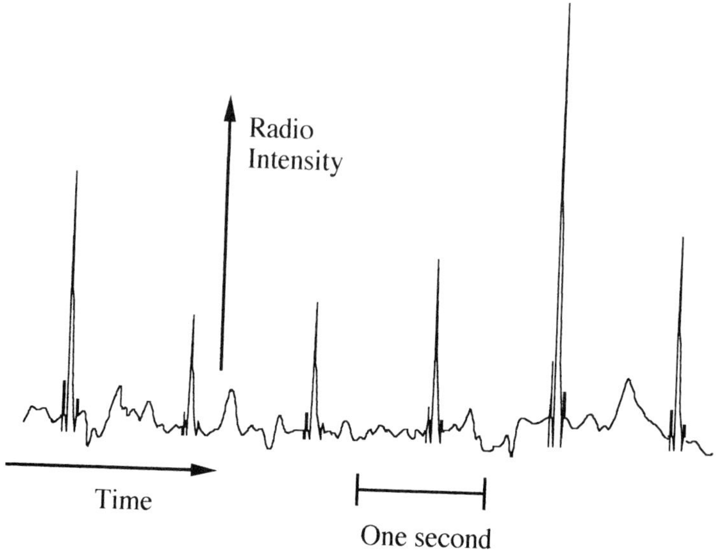

Figure 3. A typical radio signal from a pulsar.

(Figure 3). For a while it was thought that this might be a signal from life forms beyond the Earth. But with the discovery of several other similar sources it became clear that it must be a natural phenomenon. The period between the pulses, from a fraction of a second to a few seconds, and their regularity, which was much better than that of a quartz watch, quickly narrowed the possible theoretical explanations down to four: a rotating white dwarf, a vibrating white dwarf, a rotating neutron star or a vibrating neutron star; in each case the period of the radio pulses being the rotational or vibrational period of the object. The objects quickly became known as 'pulsars', a contraction of 'pulsating radio source'. Even now the majority of pulsars are detected only by their radio emissions. In a few cases, though, such as the Crab and Vela Pulsars (Figure 4), the pulses are also detected in the optical and even in the X-ray region of the spectrum.

The range of periods (now known to be from a millisecond to several seconds, with an average of 0.8 seconds) eliminated vibration as a possible explanation, because this would produce more-or-less fixed periods of about 2 seconds for a white dwarf and about

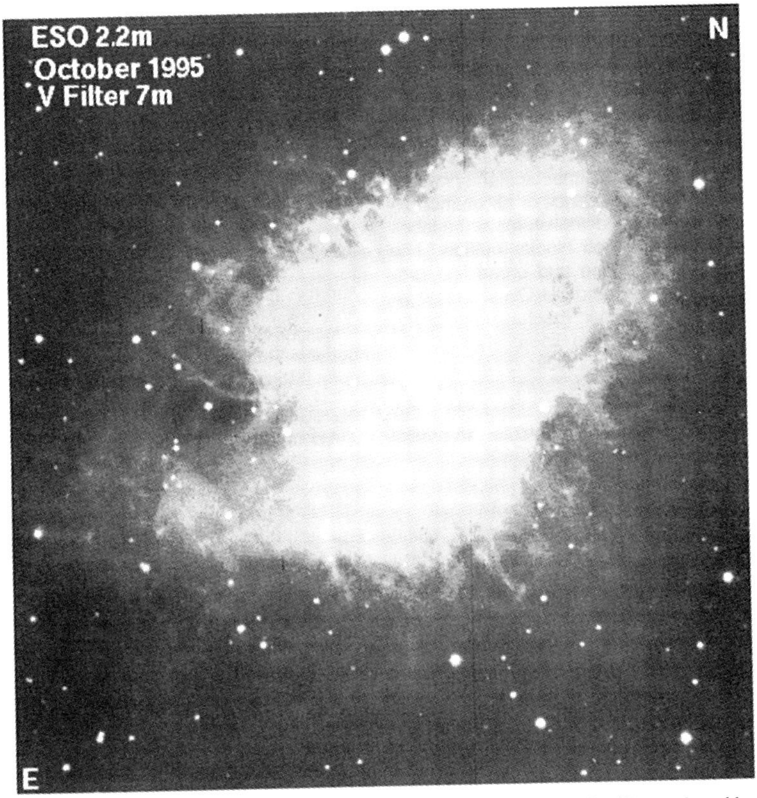

Figure 4. The Crab Pulsar (arrowed) at the heart of the Crab Nebula. (Reproduced by courtesy of the European Southern Observatory.)

0.5 ms for a neutron star. If a white dwarf were to rotate faster than about once a second, it would break up, so this could not explain the shorter-period pulsars that were observed. The only remaining explanation was therefore that these regular radio pulses were somehow produced by a rotating neutron star.

This conclusion, reached within weeks of the discovery of pulsars, is still accepted today. But despite nearly three decades of further intensive observational effort, few other definitive conclusions have been established about pulsars. The reason for the pulsed nature of the radiation is that the neutron star emits a beam of radio waves, and this swings around with the object's rotation like the beam from a lighthouse. The pulse is seen as the beam crosses the line of sight. However, the mechanism producing the radiation and the reason why it is beamed is still debated. Probably the most popular explanation is that the beam is somehow emitted along the magnetic axis of the neutron star. If this is at an angle to the rotational axis (Figure 5), then it will produce the required lighthouse effect.

The pulses can have a complex structure, and be composed of several sub-pulses. In about a third of the pulsars these sub-pulses drift with respect to each other, changing the overall pulse profile. Over the course of a few hundred pulses, though, the mean profile settles down to a very stable average shape. The origin of these sub-pulses and why they drift is still unexplained. The radiation is found to be very highly polarized, which indicates its origin as synchrotron radiation from relativistic electrons spiralling around magnetic fields. The total power going into the pulses is a small fraction of the solar luminosity. Many pulsars, however, must be losing much more energy than this in order to explain their increasing periods (see below). The rate of this energy loss can be many times the solar luminosity, and probably goes into producing very high-energy particles which the pulsar sprays off into space. Those particles may then provide the energy sources for some supernova remnants such as the Crab Nebula, and may also contribute to primary cosmic rays.

Further observation of pulsars has established some of their properties. There are now well over 600 pulsars known. The constancy of their periods can be as high as one part in 10^{16}, with one to 10^{15} being typical.[3] Many pulsars, though, do have a slow increase in their periods with time: they are slowing down. The reason for this is that they are losing angular momentum[4] as particles are flung off from their surfaces into space. The particles are lost from the equatorial regions of the neutron star, and so carry with them a disproportionate amount of angular momentum, leaving the remaining object rotating more slowly. Sometimes the slow increase in period is interrupted by a sudden, rapid decrease in the period. The neutron star has thus suddenly speeded up. These sudden changes are called *glitches* (Figure 6), and are observed especially amongst the younger pulsars. After a glitch the rotational period resumes its increase, more rapidly than normal to begin with, but soon slowing back to the old rate of change.

The explanation for the glitches is thought to be starquakes. The

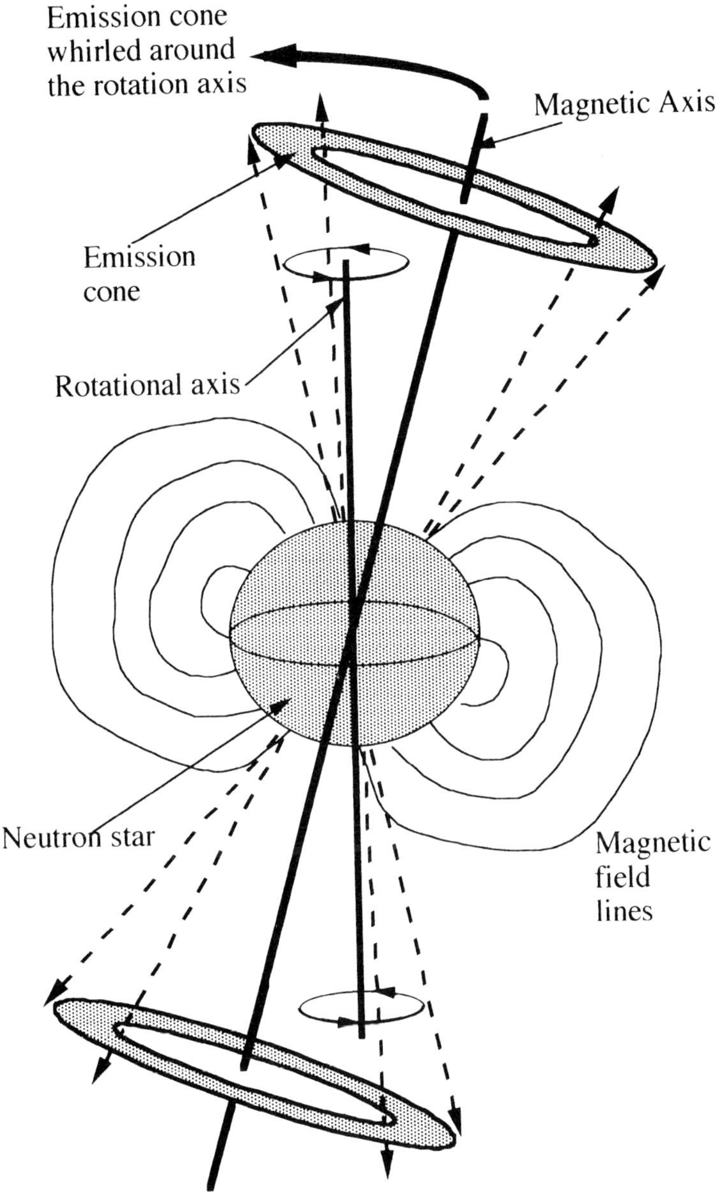

Figure 5. A widely accepted model for pulsars.

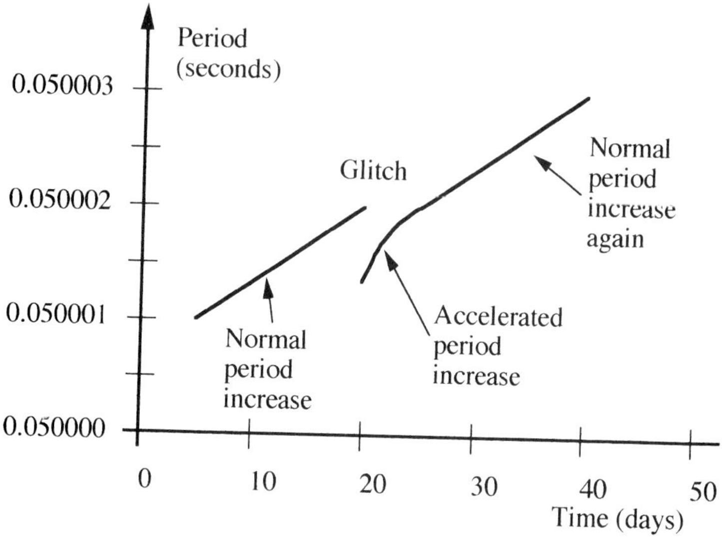

Figure 6. The variation of pulse period with time for a pulsar, together with a glitch.

neutron star will have a thin crust, probably mostly formed from iron nuclei. As the rotation slows, centrifugal 'forces' in that crust will lessen. The crust will therefore experience an increasing imbalance between gravity and centrifugal 'force'. Eventually the crust will crack under the strain, and collapse slightly, allowing it to rotate more rapidly and so get the forces acting on it back into balance. The crust of a neutron star is only very loosely connected to the underlying material, which has the form of a neutron superfluid. So immediately after a glitch, only the crust will have speeded up. The interior of the neutron star will still be rotating at the old rate. The pulsar, however, possesses a very intense magnetic field, and also a few per cent of its particles are protons and electrons, not neutrons. The electrically charged protons and electrons therefore interact with the magnetic field to bring the interior and crust back to the same period of rotation. The interior speeds up slightly during this process, while the crust, which produces the observable phenomenon of the pulsar, slows down more rapidly than normal.

How are the neutron stars which we see as pulsars produced? The answer seems to be 'in the centres of supernova explosions'. The theories of supernovæ suggest that their interiors undergo catastrophic collapses before the observed explosions. Their cores in such cases may be left as neutron stars. We therefore see with some supernova remnants such as the Crab Nebula and the Vela supernova remnant, a pulsar inside the nebula (Figure 4). Only a small proportion of pulsars (about 1%) are associated with supernova remnants, but this is not unexpected given that the lifetimes of pulsars are perhaps a hundred times those of the nebulæ. More puzzlingly, only a small proportion of supernova remnants have pulsars, so either most supernovæ do not produce neutron stars, or most neutron stars are not seen as pulsars.

It is also possible that two white dwarfs in a close binary system could merge to produce a neutron star as tidal interactions force their orbits to shrink. Such a process would probably look like a supernova to an outside observer. However, the pattern of gravitational radiation would distinguish such an object from a normal supernova produced by the explosion of a massive star. The next generation of gravity wave detectors, due to start operating in the next few years, may well be able to pick up such events.

Immediately upon formation, the neutron star will have a quite incredibly high temperature; perhaps over 10^{11} K. At this stage it is still hidden inside the supernova, but were it unshielded then, despite its small size, it would be radiating about 10^{46} W – as much energy in a tenth of a second as the Sun will radiate over its entire lifetime, or an absolute bolometric magnitude of −44 if all the energy were in the form of electro-magnetic radiation. In fact most of this energy is lost in the form of neutrinos,[5] not photons. The neutron star cools to 'only' about 10^9 K in a day, but the neutrinos continue to be the main means of cooling the neutron star for perhaps a thousand years, by which time the surface temperature will have fallen to about a million degrees. The neutron star will then have quite a complex structure. There will be a crust of heavy nuclei such as iron some hundreds of metres thick. Underlying this will be a region about half a kilometre thick containing a mixture of free neutrons and even heavier nuclei. The main part of the star will be neutrons, with a few per cent protons and electrons. Much of this will be a frictionless superfluid, because the neutrons will pair up to form bosons.[6] Finally, there may be a core where the density reaches 10^{18} kg/m^3 (four times the density of an atomic nucleus), and the constituents are pions and other elementary particles.

The collapse to an object only a few kilometres across leaves the neutron star rotating extremely rapidly, and with a very intense magnetic field. Both the angular momentum and the magnetic field will be conserved during the collapse, so that rotation periods of a few milliseconds can be expected, fitting in well with the observed periods of pulsars. The magnetic field will be compressed by a factor of 10^4 or more, leading in extreme cases to magnetic field strengths of 10^{10} T (10^{14} times the strength of the Earth's magnetic field).

Since the discovery of pulsars, neutron stars have been found in two other types of object. The first of these are the binary X-ray sources such as Hercules X-1. The X-rays from these objects vary with periods ranging from a fraction of a second to ten minutes or so. They also show signs of being members of binary systems. In Hercules X-1, for example, the X-ray pulses disappear every 41 hours for about 6 hours as the X-ray source goes into eclipse behind its companion. The X-rays are thought to be produced as material wrenched out of the companion star accretes on to the regions around the magnetic poles of a neutron star. The period of the X-ray pulses is then just the rotation period of the neutron star.

If the magnetic field of the neutron star is weaker than in the systems producing the binary X-ray stars, then the third manifestation of neutron stars may result. This is in the form of the X-ray bursters. These sources emit intense, sharp bursts of X-rays at intervals mostly in the region of a few hours, and also are binary systems. It is thought that the bursts are powered by thermonuclear runaway reactions occurring in the surface layers of the neutron star. Material is accreting on to the neutron star from its companion, but the weaker magnetic field means that the material accumulates over the whole surface and not just at the magnetic poles. The material from the companion star is largely hydrogen, and over a few hours it piles up to a sufficient depth for nuclear reactions to start converting the hydrogen to helium and heavier elements. The reactions occur explosively and produce the observed X-ray bursts. Gamma-ray bursters may also arise from neutron stars, but the evidence for this is much less conclusive. For some unknown reason a high proportion of the X-ray bursters are found close to the centres of globular clusters.

Many of the pulsars with the shortest periods, the millisecond pulsars, are also found in globular clusters. Most pulsars are isolated objects, but the high stellar density[7] in globular clusters makes pulsars there more likely to be members of binary systems. In such a binary a slowly rotating pulsar can be speeded up by tidal interactions and mass exchange with its companion to upwards of a hundred revolutions per second. Some binary X-ray sources also vary on a time scale of a millisecond, but without the stability shown by true pulses. It is possible that such variations arise from instabilities in the accretion stream from the companion.

Three specific systems (perhaps) containing neutron stars are worth mentioning in a little more detail: Geminga, SS433 and PSR 1913+16 (also known as the binary pulsar). Geminga is a strong gamma-ray and X-ray source which pulses about four times a second. Although rather different from other pulsars, since no pulses are detected in the radio region, at 100 to 150 pc away from us it is one of the nearest neutron stars to the Earth. SS433 is a highly intriguing system in which jets of material are being blasted out in opposite directions at about a quarter of the speed of light. It is a binary system, and is thought to contain a white dwarf being wrenched apart by the tides from a companion which may be a neutron star or a black hole. The jets are emitted along the rotational axis of the neutron star or black hole. The binary pulsar probably contains two neutron stars in orbit around each other. Since the pulsar is in effect a very accurate clock embedded in an intense gravitational field, the system provides a unique means of verifying the predictions of general relativity. For example the relativistic rotation of the orbit is over 30,000 times more rapid than that for Mercury.

The high accuracy of the pulsar 'clocks' has enabled the presence of planets around some pulsars to be detected. The planet's gravitational fields cause the pulsars to move very slightly as they orbit around it. Those movements result in periodic delays in the pulse arrival times at the Earth, enabling the planets to be detected. The planets are roughly comparable with the Earth in size.

What would life be like on a pulsar planet? Any life forms on it would have to be very tough. Firstly they would have to have survived being inside a supernova, and also the intense X-rays and gamma-rays emitted by the neutron star in the first few days of its life. After a few years, though, things would cool down to perhaps somewhere around the conditions presently found on Mars or the Galilean satellites of Jupiter. But the life forms would still be living in conditions close to those inside a solar flare because they would be continuously bombarded by around 10^{24} electrons, protons and neutrons per second travelling at close to the speed of light and which have been flung out from the pulsar.

What about life on neutron stars themselves? 'Ridiculous,' I hear you cry. But consider; oxygen, carbon, nitrogen and hydrogen, the basic building blocks of our type of life, will be present in the neutron star's crust. Furthermore, as it slowly cools down, it will be spending some thousands of millions of years at a temperature of around 20°C. So an ageing pulsar could be a very comfortable spot if you happen to be a 'little green man' with the shape and thickness of a sheet of tissue paper!

Finally, is there anything beyond a neutron star? Most theories suggest not. If mass is added to a neutron star until it starts to collapse, then in fractions of a second it will end up as a black hole. But one recent idea is that there could be another stable stage before black holes are reached. This would be in the form of a Strange Star. The name is highly appropriate, but it actually arises because it would be composed of up, down and strange quarks,[8] and not from the properties of the star itself. If such strange stars do exist, then they could rotate three or four times faster than a neutron star. So if ever a pulsar with a period of about half a millisecond is discovered, it is likely to be evidence of this next (and final?) stage before a black hole.

Notes

1. Baryons are a class of subatomic particles which includes protons and neutrons.
2. This is known as a Schwarzschild black hole. Rotating black holes, a more likely possibility, are known as Kerr black holes.
3. That is to say, the difference in the length of time between one pair of pulses and the next pair is only 10^{-15} (or 10^{-16}) of the interval between the pulses. For comparison, a quartz watch has a constancy of one part in 10^5.
4. Angular momentum is the rotational equivalent of momentum.
5. Subatomic particles with neutral electric charge and zero or very small rest mass. They interact very poorly indeed with matter, and so are able to fly out from the centre of even a neutron star almost without restriction.
6. Bosons are not restricted by the quantum energy requirements which led to degenerate electrons, but can all crowd into the lowest energy state.
7. In the centre of some clusters the average distance between stars can be as little as 70 AU, only twice the distance of Pluto from the Sun.
8. Quarks are sub-subatomic particles. Some of them are the building blocks of subatomic particles such as protons and neutrons. They come in six varieties named in the whimsical fashion of nuclear physicists: up, down, truth, beauty, charm and strange.

Experiments in Visual Supernova Hunting with a Large Telescope

ROBERT EVANS

In the last fifteen years, hunting for supernova explosions in other galaxies has become a major recognized way for amateur astronomers to contribute to their science. Before that time, while a few discoveries had been made, more or less by accident, it was believed in many professional circles that a large-scale systematic search by amateurs was beyond their capacity. But, beginning with the visual discovery of SN 1981A in NGC 1532, and of SN 1981D in NGC 1316, a steady trickle of amateur discoveries has been achieved. In the last few years, the trickle has become a stream, with eighteen amateur discoveries in 1996, using several types of equipment.

Also, the method of searching has changed. The first two supernovæ to be found by an amateur were found by photography. Guilio Romano of Italy found two supernovæ by this means in 1957 and 1961, before he became a professional and joined the search at Asiago Observatory. Two visual discoveries were then made by Jack Bennett of South Africa in 1968, and by Gus Johnson of the USA in 1979, using 125-mm and 200-mm telescopes respectively. Since then, the majority of amateur discoveries have been made visually. A small number have been made photographically, but CCDs are now being used by increasing numbers of amateurs, and this is being reflected in a sudden rise in the tally of supernovæ being found by amateurs who have invested in this kind of equipment.

To search for supernovæ visually, a relatively simple telescope is required, though a dark site is needed, as well as sufficient aperture to reveal stars of magnitude 14 or fainter. My early discoveries were made with a modest-quality 250-mm reflector. The supernovæ found with it were mainly of magnitude 13 or 14; the faintest discovery was just below magnitude 15.0. To search photographically or by using CCDs, the required equipment is much more expensive. A much better telescope and mounting are needed, as well as auxiliary equipment such as cameras and film, or

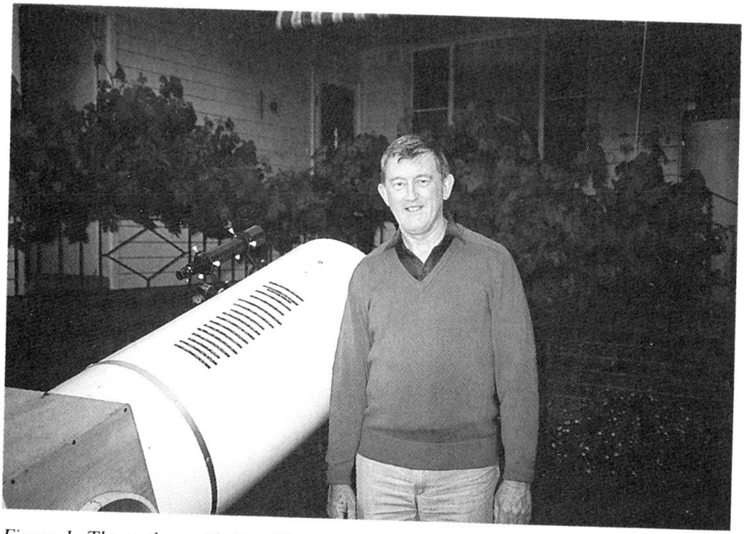

Figure 1. The author with his 410-mm telescope. Each notch on the tube represents a supernova discovery made visually with the instrument since 1986.

CCDs and fast computers. For CCD work, a dark site is not so necessary as it is for visual work or photography. At the end of 1985 I changed from using the old 250-mm telescope, and graduated to using a 410-mm equatorial Meade DS16, which had been funded for me through the CSIRO in Australia (Figure 1).

In the past decade, the competition for amateur discoveries has been in the form of professional supernova searches using fully computerized telescopes and CCDs – fully automated searches. The first group to be successful in this way was from the Astrophysics Department of the University of California at Berkeley. This group used a 0.76-m telescope, a CCD detector, and five PCs or workstations. By mid-1991 they had spent over US$5 million on the project, discovering seventeen supernovæ, and seeing three others as part of their galaxy patrol: a total of twenty. During the same six-and-a-half-year period over which the Berkeley group operated their search, I discovered twelve supernovæ with my backyard telescopes, and saw five others as part of my galaxy patrol: a total of seventeen. The small difference (three) between the results of these two efforts prompted me to publish a paper in 1994 in which the question was raised as to whether a visual searcher who

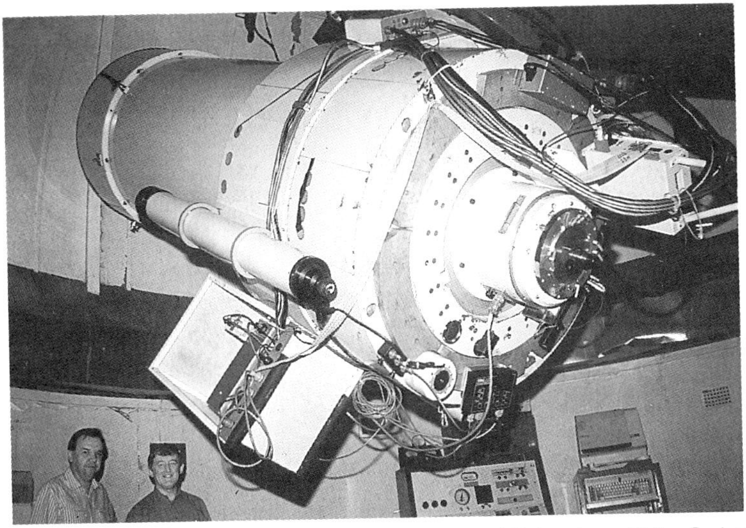

Figure 2. The 1-metre telescope of the Australian National University, at Siding Spring Observatory. (Photograph by John Shobbrook.)

could use a 0.76-m telescope for as much telescope time as the Berkeley group had enjoyed might not have produced similar or even better results.

In order to test these ideas, in 1994 I applied for government funding for an amateur-quality 1-metre telescope. For various reasons this attempt did not succeed, but it was followed by applications to use the Australian National University's 1-metre telescope at Siding Spring Observatory for four or five nights per month, preferably around new Moon, for a two-year period. Apart from wanting to make a better comparison between visual and CCD searches, nobody had previously made a serious attempt to find out what could be done with such a large telescope in this area of visual supernova searching.

The ANU's 1-metre telescope (Figure 2) was made by Boller & Chivens, is now thirty years old, and optically is of the Ritchey–Chrétien design, operating at *f*/8. The telescope tube is always kept on the same side of the mounting, and the floor is raised and lowered as needed in order to keep the observer close to the eyepiece, which is behind the main mirror. Normally, CCDs are attached, linked to computers in the nearby control room, though the telescope and dome have to be moved to the targets through controls on the observatory floor. This telescope is used by ANU graduate students and staff, and other professional researchers, almost every night of the year. So the Time Allocation Committee was generous in giving us use of the telescope for more than eighty nights from January 1995 until July 1996. This was particularly noteworthy because it was a visual project, which is extremely unusual for a professional telescope these days, and the project was staffed totally by amateur astronomers.

To use the telescope efficiently, apart from the observer, one person is needed to operate the controls, where some lighting is necessary to read galaxy catalogue positions, press buttons and follow the read-out on a screen of the telescope's actual orientation. In this way, the observer can protect his night vision as he follows the eyepiece. Several local amateur astronomers were very helpful with the sometimes tedious operating work, including Tom Cragg and John Shobbrook, while Queensland teenager Samantha Beaman often travelled down (a round trip of 1300 km) to share the observing.

Naturally we had some bad weather (about half the time), and had a good share of strong moonlight, although some of the 'dark' nights we had were magnificent, with a limiting magnitude of 17.5, or even fainter. We got wonderful views of the brighter galaxies, and saw many fainter ones we would not have seen at all through typical amateur telescopes.

The observations reported here ran from December 26, 1994, to mid-May 1996, and for three extra nights at the end of July. The scheduled nights on the telescope were scattered irregularly through that period, except for three months between March 8 and June 8, 1995, when we had no scheduled nights. 1997 will see a renewal of the observing for part of the year. About 390 usable hours yielded just over 10,000 galaxy observations, at an average rate of about 26 per hour. Three supernovæ were discovered by this means, and seven others were seen as part of the normal galaxy patrol. The discoveries were:

SN 1995G in NGC 1643 (Figure 6)
SN 1995V in NGC 1087 (Figure 3)
SN 1996al in NGC 7689 (Figure 4)

During this same nineteen-month period, my 410-mm backyard telescope, now on a cheap altazimuth mounting, performed a

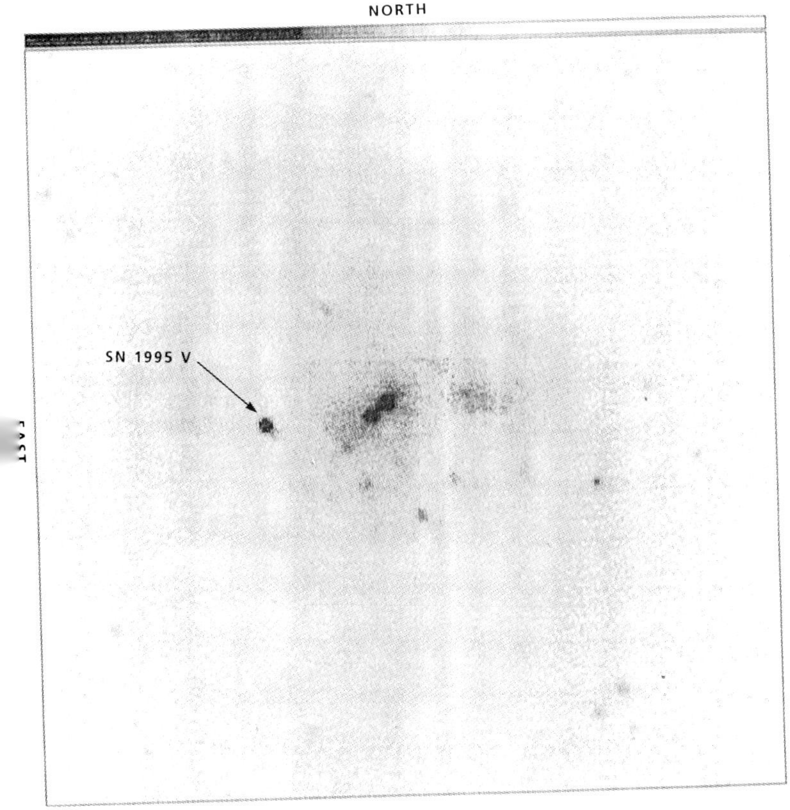

Figure 3. SN 1995V in NGC 1087. Print-out from CCD picture made to confirm the discovery by M. Dopita and C. Trung Hua using the ANU 2.3 metre telescope at Siding Spring Observatory.

purely secondary rôle, observing galaxies that were not covered by the bigger telescope because of time constraints (bad weather, or not enough scheduled time). Within this framework, 13,267 galaxy observations were made in 231 hours of observing, at an average of 57.4 observations per hour. The result of this work was that two other supernovæ were discovered, plus four others also seen with the larger telescope. These discoveries were:

SN 1995ad in NGC 2139
SN 1996X in NGC 5061 (Figure 8)

The number of different galaxies involved in this searching with both telescopes would approach 2000. These galaxies cover many types, magnitudes and sizes, regardless of what rate supernovæ might be expected to appear in any of them.

Reflections

These two searches have raised some results which I found very interesting.

1) The larger telescope certainly brings many more supernovæ within the range of observation, increasing the chances of success. But the telescope is slower to operate. As a result, fewer objects can be observed in the time available, thus reducing the possibilities of success. Clearly, a dedicated telescope having a design with easier access to the eyepiece, and with speedy location of target galaxies, would be most desirable.
2) A simple backyard telescope of about 400 mm aperture is probably the better way to find the brighter supernovæ in the closer galaxies, because these galaxies are generally spread out over many parts of the sky. With a highly manoeuvrable telescope it is easier to move from one to another, and supernovæ in these galaxies can be seen with a telescope of this size, except for ones which are intrinsically very faint.
3) Funding for a 0.75- or 1-metre telescope which can be devoted for sufficient time to this sort of searching, and which are speedy enough in operation, is a very difficult matter.

What about the Automatic Searches?

The Berkeley search is no longer functioning, but several other much smaller automatic projects are now in progress in different parts of the world. The only one in the last few years for which

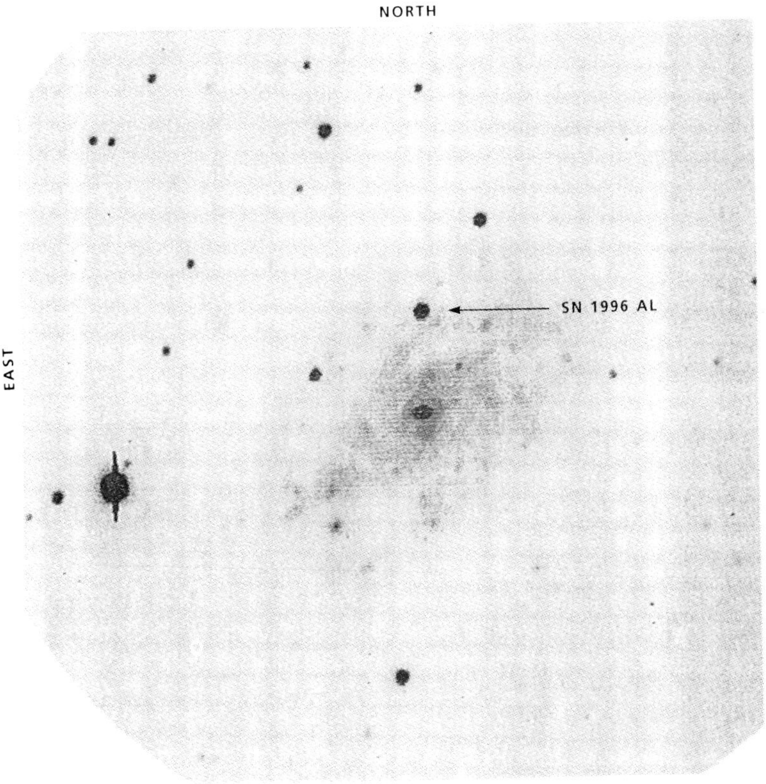

Figure 4. SN 1996al in NGC 7689. Print-out from CCD picture in visual light by Lisa Germany using the ANU 2.3 metre telescope at Siding Spring Observatory.

figures have been published, and for which any kind of comparison is possible, is the search at Perth Observatory which commenced early in 1993. A 0.6-m telescope is used, targeting 300 specially chosen large face-on spiral galaxies in which a high supernova rate is expected. Up to July 1996, 5530 observations had been made, at an average speed of 15 observations per hour, with a magnitude limit of 16.5. Five supernovæ have been discovered over a three-year period, although one other was found which could not be independently confirmed before it disappeared behind the Sun. No other supernovæ were seen.

While the efficiency of the Perth search could be improved upon with better CCDs and faster computers, it does reflect a feature which is true for amateur CCD searches also, namely that this form of searching generally takes much more time than an experienced visual observer needs to monitor the same number of galaxies. The advantages of CCD use are found in other areas, such as ease in making permanent records of what is seen, and ability to observe well from relatively light-polluted sites. As with many areas of astronomy, compromise is needed in deciding which path is to be followed. Each method of searching has its own advantages and disadvantages. No one method is going to put the others out of business. Perhaps, more than anything, what we do will be governed by what equipment we have available, and what we can afford. Certainly, visual searching has the advantage that cheaper equipment can do excellent work, although the more expensive items are now becoming more affordable to amateurs than was the case only a few years ago.

How to Conduct a Search for Supernovæ

The basic technique is to make a list of galaxies which are to be searched, and then to search them on a regular basis, preferably once a month or more often, looking for new stars.

1) Visual searching requires a telescope which reveals magnitude 14 or fainter, a dark site, good reference pictures of all the galaxies on the list, sky atlases and galaxy catalogues, and contact persons who are willing and able to verify any suspects you find.
2) Photographic searching has the same requirements, except that the telescope and mounting have to be good enough for taking good-quality time-exposure photographs regularly.

3) CCD searching, whether fully automated or not, has the same requirements, except that good computers and CCDs are also needed, along with the ability to locate galaxies to high accuracy using the computer. A dark site and a moonless night are not so essential for CCD work.

How to Evaluate a Supernova Search

Full records must be kept of all observations, including names of all galaxies observed, along with dates, times and conservative magnitude limit estimates.

As time goes by, three lists of supernovæ need to be made:

1) Supernovæ that you discover in the galaxies on your prepared list of target galaxies.
2) Supernovæ that somebody else discovers in the galaxies on your prepared list, but which you observe also as a normal part of your search routine.
3) Other supernovæ in these galaxies that you did not see as a normal part of your observing routine.

The true value of a supernova search is found by combining lists one and two, and then comparing the resulting list with the complete list of all supernovæ. That is, compare what supernovæ you actually saw within the normal routine of your search, including your discoveries, with the list of all supernovæ that were found in these galaxies. That will give the best idea of what has been missed, and how thorough your search has been. The true value does not arise from the list of your discoveries only, because what you discover depends a great deal on what competition there is for discoveries, and this can vary markedly from week to week, and from month to month.

1996 – A Record-breaking Year for Supernova Discoveries

The total number of supernovæ discovered in 1996 is eighty-one, about twenty more than the previous highest total for a year. The high tally is due largely to extensive professional searches for extremely distant supernovæ using very large or medium telescopes, and finding supernovæ in the 22–23 magnitude range. So far as I can tell, the number of supernovæ discovered by amateurs, either in part or in whole, during 1996 was eighteen. This is also a record, considerably higher than anything achieved before.

Figure 5. NGC 1643. Normal appearance of this galaxy. Photograph taken from a red survey film exposed with the UK Schmidt Telescope. Copyright: Anglo-Australian Telescope Board.

Figure 6. SN 1995G. 4415 film showing the supernova in NGC 1643, exposed 24/2/95 with UK Schmidt Telescope by P. Cass, to confirm the discovery. Copyright: Anglo-Australian Telescope Board.

Figure 7. NGC 5061. Normal appearance of this galaxy. Photograph taken from ESO 'Quick B' survey. Copyright: ESO (European Southern Observatory).

Figure 8. SN 1996X. Computer print-out from a CCD picture taken through cloud cover by Roberto Sorio using the 1-metre telescope, Siding Spring Observatory. Photo shows NGC 5061 and SN 1996X. Near maximum light. Copyright: Mount Stromlo and Siding Spring Observatories.

Table 1. Amateur or pro–am supernova discoveries in 1996.

Supernova	*Galaxy*	*Discoverer*	*Method*
1996A	anonymous	Evans	UK Schmidt search
1996O	anonymous	Evans	UK Schmidt search
1996X	NGC 5061	Evans Takamizawa	visual photographic
1996Z	NGC 2935	Johnson[1]	CCD
1996aa	NGC 5557	Johnson[1]	CCD
1996ad	anonymous	Evans	UK Schmidt search
1996ae	NGC 5775	Vagnozzi *et al.*	CCD
1996ai	NGC 5005	Bottari	CCD
1996ak	NGC 5021	Buil	CCD
1996al	NGC 7689	Evans *et al.*	visual
1996an	NGC 1084	Aoki	CCD
1996aq	NGC 5584	Aoki	CCD
1996as	anonymous	Evans	UK Schmidt search
1996bk	NGC 5308	Pesci and Massa	visual
1996bo	NGC 673	Armstrong	CCD
1996ca	NGC 7300	Aoki	CCD
1996cb	NGC 5310	Aoki	CCD
1996cc	NGC 5673	Sasaki	photographic

[1] Two discoveries in one night.

The list of amateur (or pro–am) discoveries given in Table 1 is complete to the best of my knowledge. It has truly become a world-wide project now. Six discoveries were made from Australia, and six were made from Japan. Three were made from Italy, two from the USA, one from France and one from England. Incidentally, SN 1996X was the brightest of the supernovæ found anywhere in 1996, at magnitude 13, although several others reached magnitude 14. My congratulations to all discoverers, and my best wishes to all observers who tried hard but without success. Your turn will come soon, if you persist.

This observing programme has been possible only because of the much valued support of the Director of Mount Stromlo and Siding Spring Observatories, Professor Jeremy Mould, and his Time Allocation Committee, for extensive use of the 1-metre telescope without cost to us. Everyone involved in this programme is greatly indebted to them.

The End of the Universe

IAIN NICOLSON

Although some doubts and uncertainties remain, most cosmologists accept that the Big Bang theory – the theory that the universe began in a hot, dense explosive event a finite time ago, and has been expanding ever since – is by far the best account of the origin of the Universe that is currently available. But what of the future? Will the Universe continue to expand for ever, or does some other fate lie in store? Before looking into the far future, it may be useful to summarize the history of the Universe so far, according to the Big Bang model.

The 'standard' Big Bang model

Most cosmologists believe that the Big Bang was the origin of space, time and matter. Matter did not erupt from a point into a pre-existing space; rather, space, time and matter originated together and space has been expanding ever since, carrying matter and, eventually, galaxies with it. During the first microscopic instants of the history of the Universe, the temperatures were truly enormous and the Universe was dominated by radiation (highly energetic photons). Collisions between photons produced a wide variety of particles of matter and antimatter, including quarks, antiquarks, electrons and positrons. Because there were slightly more particles than antiparticles, most of the antiparticles were quickly annihilated by colliding with particles.

About one-millionth of a second after the beginning of time, when the temperature had dropped to just below 10^{13} K, quarks clumped together to form protons and neutrons. About a hundred seconds later, when the temperature had fallen to about 10^{9} K, nuclear reactions rapidly welded protons and neutrons together to form nuclei of helium and other light elements, such as deuterium (heavy hydrogen) and lithium. Many single protons (hydrogen nuclei) remained. Space was opaque, because photons could travel only short distances before colliding with the myriad of fast-moving free electrons.

Some 300,000 years later, when the Universe had cooled to a few thousand degrees, electrons were captured by nuclei to make complete atoms, and – because photons could then travel enormous distances without suffering collisions or being absorbed – space became transparent. After this event, which is called the 'de-coupling' of matter and radiation, or the 'last scattering' of the primordial radiation, the radiation content of the universe was free to spread through the expanding volume of space. The expansion diluted and cooled the radiation, stretching the light waves to progressively longer wavelengths until today they exist as a faint background of microwave radiation, permeating all of space. Astronomers first detected this cosmic microwave background radiation more than thirty years ago.

Some time after space became transparent, matter clumped together to form galaxies, and clusters and superclusters of galaxies – how, when and in what order we do not know. Recent deep-space images, such as those of the Hubble Deep Field, indicate that galaxies had formed by the time the Universe was as little as one-fifth of its present size and perhaps less than one-tenth of its present age (about a billion years after the Big Bang). Having formed, by whatever means, the galaxies and clusters have continued to rush away from each other ever since, sharing in the overall expansion of the Universe.

Inflation

In order to overcome a number of key problems in cosmology, many cosmologists favour the idea that very early in its history, about 10^{-35} seconds after the beginning of time, the Universe experienced a short-lived but dramatic period of accelerating expansion called inflation. During the inflationary era, all distances in the Universe, and all separations between particles, increased by a huge amount – a factor of 10^{50} or more. Inflation, if indeed it did happen, blew up the Universe to such a vast size that the presently observable part of it is only a microscopic fraction of the whole.

Will the universe expand for ever?

Gravity is slowing the rate of expansion. If gravity wins the battle, the expansion will eventually cease and the galaxies will begin to fall together: slowly at first, but ever more rapidly, the Universe will begin to contract until, finally, it will collapse into a Big Crunch. A universe that expands to a finite size then collapses is called 'closed'. If gravity cannot halt the expansion, the rate of expansion will slow down towards a steady value and the expansion will continue for ever; a universe that continues to expand at a finite rate for ever is called 'open'. A third possibility is that the speeds of the galaxies and the pull of gravity are so finely balanced that the expansion can just, but only just, continue for ever, the rate of expansion approaching ever closer to zero but not becoming zero until an infinite amount of time has passed. A universe of this kind is known as an Einstein–de Sitter, or 'flat' universe.

If the Universe on the large scale is homogeneous and isotropic – the same everywhere and in every direction – then whether or not it will expand for ever will be determined by the rate of expansion and the mean density alone. If the mean density exceeds a particular value, called the critical density, gravity will win the battle and the Universe will be closed; if it is less than critical, the Universe will be open and ever-expanding. If the actual density precisely matches the critical value, the Universe will be 'flat', sitting on the fence between the two options.

The value of the critical density depends on the value of the Hubble constant, H – the number that relates the recessional velocities of galaxies to their distances. The present value of H, denoted by H_0, is generally believed to lie somewhere between about 40 and about 85 km/s/Mpc (kilometres per second per megaparsec), although several recent results have shown signs of convergence in the range 57–73 km/s/Mpc. For a value of $H_0 = 50$ km/s/Mpc, which would imply that a galaxy at a distance of 1 megaparsec would be receding from us at a speed of 50 km/s, the critical density is about 5×10^{-27} kg/m^3, roughly equivalent to 3 hydrogen atoms per cubic metre of space.

The ratio of the actual mean density to the critical density is denoted by the symbol Ω (omega). If Ω is greater than 1 (i.e. $\Omega > 1$), the actual density is greater than critical, and the Universe is closed; if $\Omega < 1$, the Universe is open, and if $\Omega = 1$ precisely, the Universe is 'flat' and only just capable of expanding for ever.

What is the present value of Ω?

If we assume that practically all the matter in the universe is luminous, and is contained in the stars and glowing gas clouds that make up the familiar galaxies, we can estimate the masses of all the galaxies in a sufficiently large volume of space, divide by the volume, and come up with a value for the mean density. Observations of this kind indicate that luminous matter contributes only about 1% of the density required to close the Universe.

However, if most of the mass is concentrated where most of the light is, in the central parts of galaxies, then stars and gas clouds in the outer parts of spiral galaxies should be revolving more slowly than those nearer the centre, just as planets farther from the Sun revolve more slowly than those closer in. In fact, the outer parts of the disks of spiral galaxies rotate as fast as, or faster than, the inner parts. Since the speed at which a star or gas cloud revolves around the centre of a galaxy depends on its distance and on the amount of mass contained within its orbit, these observations imply that most of the mass in galaxies of this kind is contained in an extended halo rather than in the central regions. A typical spiral contains 5–10 times as much dark matter as luminous matter.

A similar argument applies to clusters of galaxies. For a cluster to persist, its gravitational self-attraction must be sufficiently great to prevent individual member galaxies from escaping. The speeds at which individual galaxies move inside clusters are such that most clusters would have dispersed long ago unless they contained a great deal more mass than is directly visible. Typical clusters and groups of galaxies require tens or even hundreds of times as much dark matter as luminous matter in order to bind themselves together.

A recent technique for measuring the amount of dark matter in a cluster is to use the phenomenon of gravitational lensing. Rays of light passing a massive body or a distribution of matter are deflected in such a way that they can produce two or more images of a distant object. If the alignment is perfect, the distant object will look like a ring; if not, it will appear as two or more distorted arcs. A cluster of galaxies acts like a giant lens that distorts the shapes of distant background galaxies into arcs. Observations of this kind probe not only the inner parts of clusters, where the brighter galaxies are located, but also the outer parts, and in some cases have revealed there to be as much as 50–100 times as much dark matter as luminous matter.

Although astronomers cannot yet claim to have found enough dark matter to be certain that the Universe is closed, the observations clearly indicate that the actual value of Ω is remarkably close to 1, considering that it could have had almost any value. It is clearly at least 0.01 (we can see that much matter directly); dark matter in individual galaxies pushes the value towards 0.1, and taking into account dark matter in clusters, the value could be between 0.1 and

1. The value of Ω is unlikely to be much greater than 1. If Ω has a value of 1.01, the expansion of the Universe will cease in about 10^{12} years; a figure of 1.1 would imply that the collapse would begin 10^{11} years from now. But had Ω been as high as 10 (assuming the currently accepted range of values of *H*) the collapse would already have occurred, and we should not be here to debate the issue.

The nature of the dark matter

On the face of it, the dark matter could consist of almost anything. For example, brown dwarfs (ultra-dim stars with masses too small to enable nuclear burning to take place), planets, planetesimals and rocks could exist in great abundance in the outer fringes of galaxies or in intergalactic space and be difficult or impossible to detect with present-day techniques. However, the abundances of the lightest chemical elements in the Universe, in particular helium-3 (light-weight helium), deuterium (heavy hydrogen) and lithium, places a very important constraint on the amount of baryonic matter (conventional matter, composed of protons and neutrons) that the Universe can contain. The amounts of those elements that were produced in, and which survived through, the period of nucleosynthesis that took place during the first few minutes after the Big Bang depends strongly on the cosmic density of baryonic matter. The measured abundances of these elements indicate that no more than about 10% of the critical density can be provided by baryonic matter. If there is enough dark matter to close the Universe then at least 90% of it must be non-baryonic – a completely different type of matter that hardly ever interacts with ordinary matter, except through the agency of gravity.

Neutrinos are one of the possible contributors to the dark matter content of the Universe. These peculiar elementary particles were originally thought to have zero rest mass (a stationary neutrino would weigh nothing at all), but more recent theories suggest that neutrinos may have finite, though tiny, masses. Because there are believed to be about a billion neutrinos for every baryon in the Universe, the average mass of a neutrino could be less than one ten-thousandth of the mass of an electron, but their combined mass could still be enough to close the Universe. Recent, very tentative measurements suggest that the masses of neutrinos are perhaps too low for them to be the dominant contributor to the mass of the Universe, but it is much too early to be sure. Because they would have been moving at or close to the speed of light relative to neighbouring baryonic matter in the very early Universe, neutrinos are an example of what has come to be known as 'hot' dark matter. Another possibility is 'cold' dark matter: slower-moving elementary particles, whose existence has been predicted by various theories of the fundamental forces of nature: particles such as ultra-lightweight 'axions', or a range of more massive particles collectively referred to as WIMPS (weakly interacting massive particles). While various experiments are under way to determine the masses (if any) of neutrinos and to attempt to detect WIMPS, no conclusive results are yet available. At the time of writing, these exotic dark matter candidates remain theoretical possibilities only.

Although generally reckoned to be 'outside contenders', black holes may yet prove to be significant contributors to the mean density. Michael Hawkins, of the Royal Observatory, Edinburgh, has suggested that certain characteristic brightenings and fadings of quasars may be the result of gravitational lensing caused by black holes comparable in mass to the planet Jupiter passing between those quasars and the Earth. Hawkins contends that so many of these events are seen that these compact objects may provide enough mass to close the Universe. Whether or not these controversial suggestions are correct, if black holes are indeed an important constituent of the dark matter then the abundances of the light elements mentioned above rule out the possibility that the majority of black holes were formed from baryonic matter during or after the time when cosmic nucleosynthesis was taking place. However, there have been suggestions that at a very early stage, perhaps as early as one-millionth of a second after the Big Bang – when quarks were starting to clump together to make protons and neutrons – baryonic matter could have been squeezed with such violence as to form black holes of about Jupiter's mass long before nucleosynthesis itself took place.

The inflationary hypothesis requires Ω to be very close, perhaps indistinguishably close, to 1. The greater the amount of inflation, the nearer the Universe comes to being flat and the closer Ω approaches the value 1. Although the observations show that Ω is within a factor of ten of being equal to 1, for the moment we simply do not know whether the Universe is open, flat or closed. The debate seems set to continue for a long time to come.

The far future in an open Universe

If the Universe is open, the expansion will go on for ever, the galaxies getting farther and farther apart with the passage of time. Individual stars will consume all their nuclear fuels and die. Most will end their lives as white dwarfs which, after billions and billions of years, will cool down to become cold black dwarfs. Some of the more massive stars will explode as supernovæ, their cores collapsing to become ultra-compact neutron stars, while the most massive stars of all will collapse to form black holes. The least massive stars will outlive the Sun many times over, but even so, after 10^{12} to 10^{14} years all nuclear reactions in every star in every galaxy will have ceased, and galaxies will have become cold dark places.

Although close encounters between stars are extremely rare events, given sufficient time, many encounters between dead stars will take place. In each encounter, one star will gain energy and the other will lose energy. Even without any encounters of this kind, an orbiting star will gradually lose energy by radiating gravitational waves and so, very slowly, will migrate closer to the centre of its galaxy. Close encounters will accelerate this process. Over extremely long periods of time, most dead stars will evaporate from their host galaxies and the remainder will coalesce into gigantic 'galactic' black holes at their centres. A similar process is likely to happen to clusters and superclusters of galaxies, with dead galaxies merging at their centres to form 'supergalactic' black holes, and others being ejected into intercluster space.

After about 10^{20} years (ten billion times longer than the present age of the Universe), the Universe is likely to consist of gigantic black holes, individual dead stars, planets and rocks, gas, elementary particles (such as electrons and neutrinos) and low-energy photons spread ever more thinly throughout the ever-increasing volume of space. By this time, the average distance between galactic black holes will be about 10^{12} parsecs – about a hundred times larger than the radius of the presently-observable Universe.

What happens next depends on whether or not protons are stable, everlasting particles. Current theories suggest that protons, the basic building blocks of atomic nuclei, eventually decay into lighter particles and radiation. Typically, such an event would involve a proton decaying into a positron (an anti-electron) and a neutral pi-meson, or pion. The pion, being highly unstable, would almost immediately decay into photons so that the net result of proton decay would be to produce photons and a positron. The average lifetime of a proton is likely to be well in excess of 10^{31} years, since experiments aimed at detecting proton decay have so far failed to yield a single unequivocal detection. Nevertheless, if protons do decay, then in, say, 10^{33} years' time all conventional matter will have disintegrated, predominantly into positrons and radiation. Dead stars, planets and rocks alike will have vanished from the Universe to be replaced by positrons and photons to add to the existing 'sea' of electrons, photons and neutrinos.

Black holes will still be around at this stage but, if Stephen Hawking's ideas are correct, even they will not last for ever. The uncertainty principle of quantum physics permits pairs of particles and antiparticles to form, collide again and annihilate in 'empty' space; such particles are called 'virtual' because they vanish again before they can be observed (though certain side-effects of their ephemeral existence can be observed). According to Hawking, when this process happens in the powerful gravitational field close to the event horizon (boundary) of a black hole, one member of the pair may fall into the hole, leaving the other with no partner with which to annihilate. Although it, too, may fall into the hole, if it has sufficient energy it will escape instead. By this mechanism (called 'Hawking radiation') the black hole will radiate photons and particles like a conventional 'hot' body.

This emission effectively removes energy from the black hole's gravitational field and so reduces its mass. The smaller it gets, the faster it 'leaks', so that eventually a black hole will explode in a burst of particles and radiation. It will take at least 10^{66} years for a stellar-mass black hole to evaporate in this way; indeed, for a very long time to come – roughly 10^{22} years – the net inflow of matter into a typical black hole will far exceed the losses by Hawking radiation. Depending on its mass, a galactic black hole would take about 10^{96} years, and a supergalactic black hole about 10^{105} years, to evaporate.

Even if proton decay were not to happen in the way previously described, other quantum processes – exceedingly rare but inevitable given sufficiently long time scales – would result in the ultimate decay of matter. In essence, the process hinges on a phenomenon called quantum-mechanical tunnelling whereby a particle which, in 'classical' (non-quantum) terms, has insufficient energy to surmount an energy barrier, nevertheless has a small but finite chance of penetrating that barrier. By this means, white dwarfs would eventually, over unimaginably long time scales, 'decay' into neutron stars and then into black holes, which would subsequently decay by the Hawking process into particles and radiation. Depending on whether or not there is a fundamental minimum mass for a black hole, even planets, rocks and individual protons could decay by a similar basic mechanism. However long it takes, and whatever the exact details of the processes involved, the end-result will be that all matter will end up as an ever more dilute mixture of electrons, positrons, neutrinos and photons.

The degree of dilution of matter at these times is almost impossible to visualize. In 10^{66} years' time, the average distance between a typical electron and a typical positron will be about 10^{12} pc (over a hundred thousand times the radius of the presently observed Universe). Despite these inconceivable distances, electrons and positrons – by their mutual attractions – would eventually form pairs and would begin to orbit each other, forming 'atoms' of 'positronium'. Gradually, the orbiting pair would lose energy, slowly spiralling together until they collide, annihilate, and turn into photons. This process has been estimated to take about 10^{116} years.

Some electrons and positrons will avoid this fate, so that ultimately the Universe will consist of inconceivably widely separated photons, neutrinos and occasional electrons and positrons, and will progressively become ever darker and colder. The same basic scenario will also apply to the far future of the 'flat' universe. Nothing else seems likely ever to happen in this dismal long-range scenario – unless, of course, advanced intelligent life forms can change things, at least locally.

The fate of a closed Universe

If the Universe is closed, it will expand to a finite size then collapse. If the expansion ceases in the not-too-distant future – say in 10^{11} to 10^{12} years' time – some stars will still be shining within their host galaxies, and there may well be astronomers around to witness what happens next. The redshifts that were previously visible in the spectra of galaxies will change to blueshifts as the galaxies begin to approach each other. Because light takes time to travel from distant galaxies, the contraction will already have begun before observers start to see blueshifts – first in nearby galaxies, and later in more distant ones. For a long time, nearby galaxies will be showing blueshifts while distant ones are still displaying redshifts.

Gradually at first, and then more rapidly, the temperature of the background radiation will begin to rise. By about 10 billion years before the Big Crunch, it will have climbed back up to its present value (about 3 K).

About 100 million years before the Big Crunch, galaxies will begin to merge and lose their identities. Ten million years before 'the End', the whole of space will be warmer than the surface of present-day Earth, and our (carbon-based and water-dependent) form of life will find it difficult to survive. A hundred thousand years before the Crunch, the temperature everywhere will be about 10,000 K, hotter than the surface of the Sun. Stars – unable to radiate energy from their surfaces – will explode, and the whole Universe will become filled with an opaque 'soup' of plasma (a mixture of atomic nuclei and electrons) and radiation, similar in nature to the Big Bang fireball from which it first emerged.

With a hundred seconds to go, and with the temperature at around 10^9 K, atomic nuclei will break into their constituent protons and neutrons – the chemical elements will be destroyed. One-millionth of a second before the end, the temperature will be about 10^{13} K, and protons and neutrons will disintegrate into quarks. One-millionth of a second later, the Big Crunch will arrive. Conventional theory suggests that this will be the end of space, time and matter. Everything will be snuffed out of existence in a singularity – a state of infinite compression in which space, time and matter can no longer exist. There would be no 'after' because time itself would cease to exist in the Big Crunch.

Rebirth?

Some have suggested that, because the Big Crunch would to some extent replicate the conditions that existed in the Big Bang, the Crunch would be followed by a new Bang and a re-expansion of the Universe. According to this hypothesis, the Universe would rebound before reaching the ultimate state of infinite compression that terminates space, time and matter. It would enter a new cycle of expansion and contraction, and this cyclic process could continue indefinitely. If this hypothesis is correct, then there is no reason to suppose that the present cycle is the first; there may have been countless previous cycles, and the process will go on and on into the indefinite future.

However, it must be said that we know of no force that could halt the collapse and initiate a bounce. If it were to happen, some new and presently unknown force or forces would have to come into play. Present-day theories of forces, particles, space and time cannot deal with the conditions that would prevail in the final instant of the collapse. Perhaps spacetime would break down into a kind of turbulent quantum 'foam' from which almost anything could emerge. A quantum theory of gravity is needed to handle this situation, but no such theory is available at present.

Not quite running the film backwards!

The Big Crunch, though similar, would not be exactly the same as a Big Bang in reverse. Throughout the lifetime of the Universe, stars and other processes convert matter into radiation, so that the radiation content of the Universe increases and the matter content decreases with time. If the Big Crunch does not happen until long after stars have died and protons have decayed, even more of the content of the Universe will be radiation, and there will be effectively no baryons to break down into quarks during the final rush to oblivion. Whenever the Big Crunch occurs, more radiation will be going in than came out of the Big Bang. In the present expanding Universe, light waves are stretched or redshifted by the expansion, and this reduces the energy of the photons. During the collapse phase, radiation is blueshifted and the energies of photons are increased; this will make the Universe collapse faster in its late stages and will cause it to become even hotter than the original Big Bang.

If the Universe bounces before space, time and matter are totally annihilated, the bounce should therefore be more energetic than the Big Bang that preceded it. Each successive cycle would be bigger and longer than its predecessor until, ultimately, the cycles would become so long that the inhabitants of such a universe (if any) might be unable to decide whether their universe is closed, flat or open. But if all 'memory' of the previous cycle were to be completely wiped out in the Crunch, a new cycle would be, in effect, a completely independent universe unrelated in any way to the one that preceded it. In these circumstances it would hardly be reasonable to think of the process as a continuous oscillation. Conditions could be totally different in the 'next' cycle: different forces, different laws, different values for the fundamental constants, perhaps even different numbers of dimensions. There is no guarantee that conditions in a new cycle, or a new universe, would permit the existence of life as we know it or of sentient beings of any kind. Perhaps only a small proportion of cycles, or of independent universes, are capable of supporting life.

Other universes – new universes?

If our Universe began as a tiny quantum fluctuation which inflated rapidly, creating its own space and matter as it went, there is no reason why this event should be unique. There could be any number – perhaps an infinite number – of universes, each completely independent within its own expanding bubble of spacetime, blissfully unaware of the others.

It has even been suggested that quantum effects could create new expanding spacetime bubbles in our own Universe. Rather than expanding through our Universe with alarming consequences, these would branch off from our spacetime, creating their own space and time as they grew. In this scenario, even if our own Universe is fated to collapse into oblivion in a Big Crunch, it could give birth to new universes beforehand, some of which might be capable of supporting life. Perhaps advanced beings in the far future might even be able to concentrate enough energy in a small enough volume of space to make a new universe and somehow escape into it, avoiding the Big Crunch. Following this line of argument to its logical conclusion, cosmologist Edward Harrison has speculated that our own Universe may have been deliberately created in this way. Such ideas are, of course, sheer speculation, and may turn out to bear no relationship whatever to reality.

'The End' as of now

At present we do not know what fate lies in store for the Universe. We do not know if it is destined to expand for ever, to collapse or to perch uneasily on the fence. There is little chance that the problem will be resolved at least until the values of the Hubble constant, the mean cosmic density and the dark matter content of the Universe have been determined much more precisely. If the Universe is indeed open, it will expand for ever, becoming colder, darker and ever more dilute as it evolves. After truly colossal intervals of time ordinary matter (atoms and nuclei) may disappear completely from the Universe and the era of stars and galaxies will turn out to have been no more than a brief blip in the history of a Universe that is headed towards a dark featureless eternity. If the Universe is closed, it is fated to collapse – sooner or later – into a Big Crunch. So far as we know, the Big Crunch will be, literally, the end of space, time and matter, although the speculative concept of an oscillating universe, or of new universes, remains a tantalizing prospect for those who find unpalatable the bleak emptiness of the open universe or the final annihilation of the closed one. But if inflation did indeed take place, the Universe may be so close to sitting on the fence that we may never know its ultimate fate.

Part III: Miscellaneous

Some Interesting Variable Stars

JOHN ISLES

The following stars are of interest for many reasons. Of course, the periods and ranges of many variables are not constant from one cycle to another. Finder charts are given on the pages following this list for those stars marked with an asterisk.

Star	*RA* h	m	*Declination* °	′	*Range*	*Type*	*Period* days	*Spectrum*
R Andromedæ	00	24.0	+38	35	5.8–14.9	Mira	409	S
W Andromedæ	02	17.6	+44	18	6.7–14.6	Mira	396	S
U Antliæ	10	35.2	−39	34	5–6	Irregular	–	C
Theta Apodis	14	05.3	−76	48	5–7	Semi-regular	119	M
R Aquarii	23	43.8	−15	17	5.8–12.4	Symbiotic	387	M+Pec
T Aquarii	20	49.9	−05	09	7.2–14.2	Mira	202	M
R Aquilæ	19	06.4	+08	14	5.5–12.0	Mira	284	M
V Aquilæ	19	04.4	−05	41	6.6– 8.4	Semi-regular	353	C
Eta Aquilæ	19	52.5	+01	00	3.5– 4.4	Cepheid	7.2	F–G
U Aræ	17	53.6	−51	41	7.7–14.1	Mira	225	M
R Arietis	02	16.1	+25	03	7.4–13.7	Mira	187	M
U Arietis	03	11.0	+14	48	7.2–15.2	Mira	371	M
R Aurigæ	05	17.3	+53	35	6.7–13.9	Mira	458	M
Epsilon Aurigæ	05	02.0	+43	49	2.9– 3.8	Algol	9892	F+B
R Boötis	14	37.2	+26	44	6.2–13.1	Mira	223	M
W Boötis	14	43.4	+26	32	4.7– 5.4	Semi-regular?	450?	M
X Camelopardalis	04	45.7	+75	06	7.4–14.2	Mira	144	K–M
R Cancri	08	16.6	+11	44	6.1–11.8	Mira	362	M
X Cancri	08	55.4	+17	14	5.6– 7.5	Semi-regular	195?	C
*FW Canis Majoris	07	24.7	−16	12	5.0– 5.5	Gamma Cas	–	–
*R Canis Majoris	07	19.5	−16	24	5.7– 6.3	Algol	1.1	F
S Canis Minoris	07	32.7	+08	19	6.6–13.2	Mira	333	M
*VY Canis Majoris	07	23.0	−25	46	6.5– 9.5	Unique?	–	–
R Canum Ven.	13	49.0	+39	33	6.5–12.9	Mira	329	M
R Carinæ	09	32.2	−62	47	3.9–10.5	Mira	309	M
S Carinæ	10	09.4	−61	33	4.5– 9.9	Mira	149	K–M
l Carinæ	09	45.2	−62	30	3.3– 4.2	Cepheid	35.5	F–K
Eta Carinæ	10	45.1	−59	41	−0.8– 7.9	Irregular	–	Pec
R Cassiopeiæ	23	58.4	+51	24	4.7–13.5	Mira	430	M
S Cassiopeiæ	01	19.7	+72	37	7.9–16.1	Mira	612	S
W Cassiopeiæ	00	54.9	+58	34	7.8–12.5	Mira	406	C
Gamma Cass.	00	56.7	+60	43	1.6– 3.0	Irregular	–	B
Rho Cassiopeiæ	23	54.4	+57	30	4.1– 6.2	Semi-regular	–	F–K
R Centauri	14	16.6	−59	55	5.3–11.8	Mira	546	M
S Centauri	12	24.6	−49	26	7–8	Semi-regular	65	C
T Centauri	13	41.8	−33	36	5.5– 9.0	Semi-regular	90	K–M
S Cephei	21	35.2	+78	37	7.4–12.9	Mira	487	C
T Cephei	21	09.5	+68	29	5.2–11.3	Mira	388	M
Delta Cephei	22	29.2	+58	25	3.5– 4.4	Cepheid	5.4	F–G
Mu Cephei	21	43.5	+58	47	3.4– 5.1	Semi-regular	730	M
U Ceti	02	33.7	−13	09	6.8–13.4	Mira	235	M
W Ceti	00	02.1	−14	41	7.1–14.8	Mira	351	S
*Omicron Ceti	02	19.3	−02	59	2.0–10.1	Mira	332	M

Star	*RA* h	m	*Declination* °	′	*Range*	*Type*	*Period* days	*Spectrum*
R Chamæleontis	08	21.8	−76	21	7.5–14.2	Mira	335	M
T Columbæ	05	19.3	−33	42	6.6–12.7	Mira	226	M
R Comæ Ber.	12	04.3	+18	47	7.1–14.6	Mira	363	M
R Coronæ Bor.	15	48.6	+28	09	5.7–14.8	R Coronæ Bor.	–	C
S Coronæ Bor.	15	21.4	+31	22	5.8–14.1	Mira	360	M
T Coronæ Bor.	15	59.6	+25	55	2.0–10.8	Recurrent nova	–	M+Pec
V Coronæ Bor.	15	49.5	+39	34	6.9–12.6	Mira	358	C
W Coronæ Bor.	16	15.4	+37	48	7.8–14.3	Mira	238	M
R Corvi	12	19.6	−19	15	6.7–14.4	Mira	317	M
R Crucis	12	23.6	−61	38	6.4– 7.2	Cepheid	5.8	F–G
R Cygni	19	36.8	+50	12	6.1–14.4	Mira	426	S
U Cygni	20	19.6	+47	54	5.9–12.1	Mira	463	C
W Cygni	21	36.0	+45	22	5.0– 7.6	Semi-regular	131	M
RT Cygni	19	43.6	+48	47	6.0–13.1	Mira	190	M
SS Cygni	21	42.7	+43	35	7.7–12.4	Dwarf nova	50±	K+Pec
CH Cygni	19	24.5	+50	14	5.6– 9.0	Symbiotic	–	M+B
Chi Cygni	19	50.6	+32	55	3.3–14.2	Mira	408	S
R Delphini	20	14.9	+09	05	7.6–13.8	Mira	285	M
U Delphini	20	45.5	+18	05	5.6– 7.5	Semi-regular	110?	M
EU Delphini	20	37.9	+18	16	5.8– 6.9	Semi-regular	60	M
Beta Doradûs	05	33.6	−62	29	3.5– 4.1	Cepheid	9.8	F–G
R Draconis	16	32.7	+66	45	6.7–13.2	Mira	246	M
T Eridani	03	55.2	−24	02	7.2–13.2	Mira	252	M
R Fornacis	02	29.3	−26	06	7.5–13.0	Mira	389	C
R Geminorum	07	07.4	+22	42	6.0–14.0	Mira	370	S
U Geminorum	07	55.1	+22	00	8.2–14.9	Dwarf nova	105±	Pec+M
*Zeta Geminorum	07	04.1	+20	34	3.6– 4.2	Cepheid	10.2	F–G
*Eta Geminorum	06	14.9	+22	30	3.2– 3.9	Semi-regular	233	M
S Gruis	22	26.1	−48	26	6.0–15.0	Mira	402	M
S Herculis	16	51.9	+14	56	6.4–13.8	Mira	307	M
U Herculis	16	25.8	+18	54	6.4–13.4	Mira	406	M
Alpha Herculis	17	14.6	+14	23	2.7– 4.0	Semi-regular	–	M
68, u Herculis	17	17.3	+33	06	4.7– 5.4	Algol	2.1	B+B
R Horologii	02	53.9	−49	53	4.7–14.3	Mira	408	M
U Horologii	03	52.8	−45	50	6–14	Mira	348	M
R Hydræ	13	29.7	−23	17	3.5–10.9	Mira	389	M
U Hydræ	10	37.6	−13	23	4.3– 6.5	Semi-regular	450?	C
VW Hydri	04	09.1	−71	18	8.4–14.4	Dwarf nova	27±	Pec
R Leonis	09	47.6	+11	26	4.4–11.3	Mira	310	M
R Leonis Minoris	09	45.6	+34	31	6.3–13.2	Mira	372	M
R Leporis	04	59.6	−14	48	5.5–11.7	Mira	427	C
Y Libræ	15	11.7	−06	01	7.6–14.7	Mira	276	M
RS Libræ	15	24.3	−22	55	7.0–13.0	Mira	218	M
Delta Libræ	15	01.0	−08	31	4.9– 5.9	Algol	2.3	A
R Lyncis	07	01.3	+55	20	7.2–14.3	Mira	379	S
R Lyræ	18	55.3	+43	57	3.9– 5.0	Semi-regular	46?	M
RR Lyræ	19	25.5	+42	47	7.1– 8.1	RR Lyræ	0.6	A–F
Beta Lyræ	18	50.1	+33	22	3.3– 4.4	Eclipsing	12.9	B
U Microscopii	20	29.2	−40	25	7.0–14.4	Mira	334	M
*U Monocerotis	07	30.8	−09	47	5.9– 7.8	RV Tauri	91	F–K
V Monocerotis	06	22.7	−02	12	6.0–13.9	Mira	340	M
R Normæ	15	36.0	−49	30	6.5–13.9	Mira	508	M
T Normæ	15	44.1	−54	59	6.2–13.6	Mira	241	M
R Octantis	05	26.1	−86	23	6.3–13.2	Mira	405	M
S Octantis	18	08.7	−86	48	7.2–14.0	Mira	259	M
V Ophiuchi	16	26.7	−12	26	7.3–11.6	Mira	297	C
X Ophiuchi	18	38.3	+08	50	5.9– 9.2	Mira	329	M
RS Ophiuchi	17	50.2	−06	43	4.3–12.5	Recurrent nova	–	OB+M
U Orionis	05	55.8	+20	10	4.8–13.0	Mira	368	M
W Orionis	05	05.4	+01	11	5.9– 7.7	Semi-regular	212	C
Alpha Orionis	05	55.2	+07	24	0.0– 1.3	Semi-regular	2335	M

SOME INTERESTING VARIABLE STARS

Star	*RA* *h*	*m*	*Declination* °	′	*Range*	*Type*	*Period days*	*Spectrum*
S Pavonis	19	55.2	−59	12	6.6–10.4	Semi-regular	381	M
Kappa Pavonis	18	56.9	−67	14	3.9– 4.8	Cepheid	9.1	G
R Pegasi	23	06.8	+10	33	6.9–13.8	Mira	378	M
Beta Pegasi	23	03.8	+28	05	2.3– 2.7	Irregular	–	M
X Persei	03	55.4	+31	03	6.0– 7.0	Gamma Cas	–	O9.5
Beta Persei	03	08.2	+40	57	2.1– 3.4	Algol	2.9	B
Rho Persei	03	05.2	+38	50	3.3– 4.0	Semi-regular	50?	M
Zeta Phœnicis	01	08.4	−55	15	3.9– 4.4	Algol	1.7	B+B
R Pictoris	04	46.2	−49	15	6.4–10.1	Semi-regular	171	M
L^2 Puppis	07	13.5	−44	39	2.6– 6.2	Semi-regular	141	M
*RS Puppis	08	13.1	−34	35	6.5– 7.7	Cepheid	41.4	G
T Pyxidis	09	04.7	−32	23	6.5–15.3	Recurrent nova	7000±	Pec
U Sagittæ	19	18.8	+19	37	6.5– 9.3	Algol	3.4	B+G
WZ Sagittæ	20	07.6	+17	42	7.0–15.5	Dwarf nova	11900±	A
R Sagittarii	19	16.7	−19	18	6.7–12.8	Mira	270	M
RR Sagittarii	19	55.9	−29	11	5.4–14.0	Mira	336	M
RT Sagittarii	20	17.7	−39	07	6.0–14.1	Mira	306	M
RU Sagittarii	19	58.7	−41	51	6.0–13.8	Mira	240	M
RY Sagittarii	19	16.5	−33	31	5.8–14.0	R Coronæ Bor.	–	G
RR Scorpii	16	56.6	−30	35	5.0–12.4	Mira	281	M
RS Scorpii	16	55.6	−45	06	6.2–13.0	Mira	320	M
RT Scorpii	17	03.5	−36	55	7.0–15.2	Mira	449	S
S Sculptoris	00	15.4	−32	03	5.5–13.6	Mira	363	M
R Scuti	18	47.5	−05	42	4.2– 8.6	RV Tauri	146	G–K
R Serpentis	15	50.7	+15	08	5.2–14.4	Mira	356	M
S Serpentis	15	21.7	+14	19	7.0–14.1	Mira	372	M
T Tauri	04	22.0	+19	32	9.3–13.5	Irregular	–	F–K
SU Tauri	05	49.1	+19	04	9.1–16.9	R Coronæ Bor.	–	G
Lambda Tauri	04	00.7	+12	29	3.4– 3.9	Algol	4.0	B+A
R Trianguli	02	37.0	+34	16	5.4–12.6	Mira	267	M
R Ursæ Majoris	10	44.6	+68	47	6.5–13.7	Mira	302	M
T Ursæ Majoris	12	36.4	+59	29	6.6–13.5	Mira	257	M
U Ursæ Minoris	14	17.3	+66	48	7.1–13.0	Mira	331	M
R Virginis	12	38.5	+06	59	6.1–12.1	Mira	146	M
S Virginis	13	33.0	−07	12	6.3–13.2	Mira	375	M
SS Virginis	12	25.3	+00	48	6.0– 9.6	Semi-regular	364	C
R Vulpeculæ	21	04.4	+23	49	7.0–14.3	Mira	137	M
Z Vulpeculæ	19	21.7	+25	34	7.3– 8.9	Algol	2.5	B+A

R and FW Canis Majoris

Comparison stars:

A = 4.96
B = 5.45
C = 5.46
D = 5.78
E = 6.05
F = 6.09
G = 6.6
H = 6.77

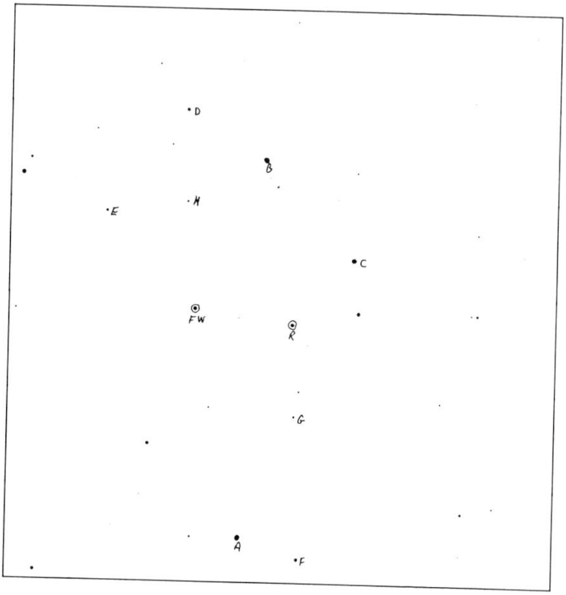

VY Canis Majoris

Comparison stars:

C = 7.0
D = 7.1
E = 8.1
F = 8.4
G = 8.8
H = 9.4

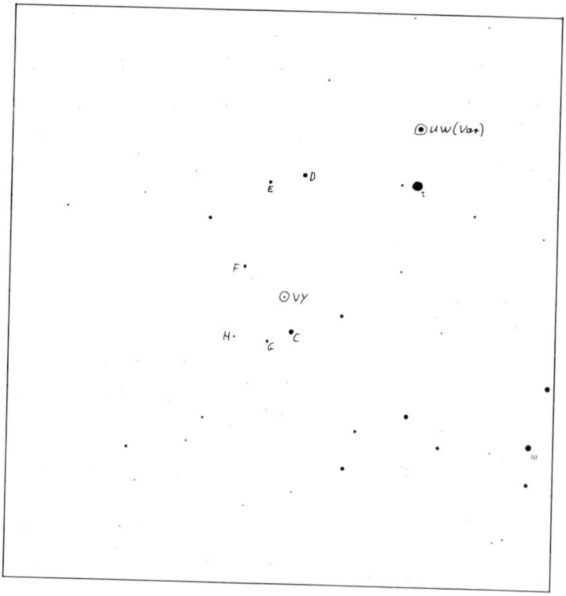

Mira

Comparison stars:

Alpha (α)	= 2.52 (off map)
Gamma (γ)	= 3.46
Delta (δ)	= 4.06
Nu (ν)	= 4.87
N	= 5.34
P	= 5.41
R	= 6.00
S	= 6.32
T	= 6.49
U	= 7.19
W	= 8.06
X	= 8.42
y	= 9.00
z	= 9.33

Eta and Zeta Geminorum

Comparison stars:

Epsilon (ε) Gem	= 2.98
Zeta (ζ) Tau	= 3.03
Xi (ξ) Gem	= 3.34
Lambda (λ) Gem	= 3.59
Nu (ν) Gem	= 4.14
1 Gem	= 4.15

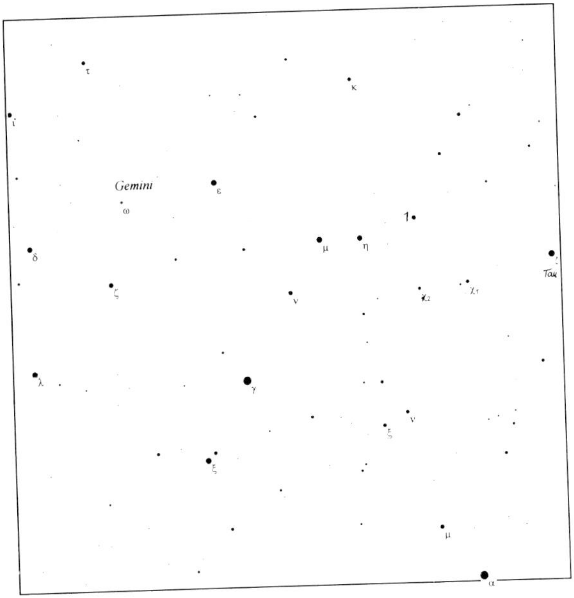

U Monocerotis

Comparison stars:

C = 5.72
D = 5.85
E = 6.00
F = 6.62
G = 6.97
H = 7.51
K = 7.81
L = 8.03

RS Puppis

Comparison stars:

A = 6.4
B = 6.4
C = 7.0
D = 7.4
E = 7.6
F = 8.2
G = 8.3

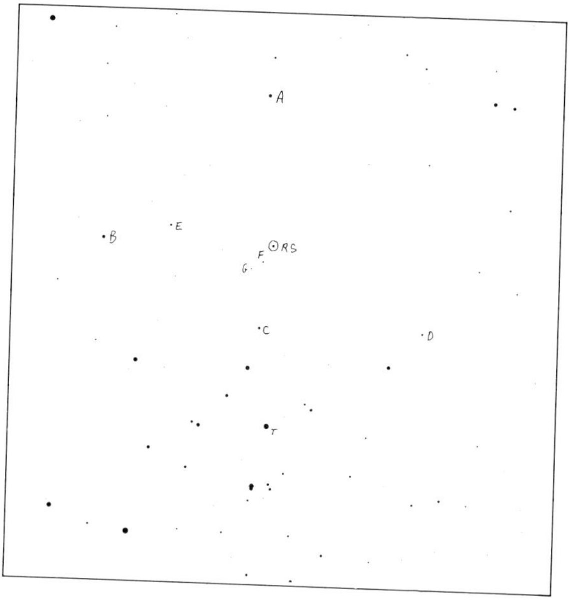

Mira Stars: Maxima, 1998

JOHN ISLES

Below are the predicted dates of maxima for Mira stars that reach magnitude 7.5 or brighter at an average maximum. Individual maxima can in some cases be brighter or fainter than average by a magnitude or more, and all dates are only approximate. The positions, extreme ranges and mean periods of these stars can be found in the preceding list of interesting variable stars.

Star	*Mean magnitude at maximum*	*Dates of maxima*
R Andromedæ	6.9	June 1
W Andromedæ	7.4	July 30
R Aquarii	6.5	May 5
R Aquilæ	6.1	July 26
R Boötis	7.2	July 18
R Cancri	6.8	Oct. 30
S Canis Minoris	7.5	Mar. 5
R Carinæ	4.6	Apr. 11
S Carinæ	5.7	May 21, Oct. 18
R Cassiopeiæ	7.0	May 11
R Centauri	5.8	Nov. 8
T Cephei	6.0	Apr. 18
U Ceti	7.5	Jan. 8, Aug. 30
Omicron Ceti	3.4	Jan. 13, Dec. 11
T Columbæ	7.5	June 7
S Coronæ Borealis	7.3	Oct. 7
V Coronæ Borealis	7.5	May 4
R Corvi	7.5	Aug. 8
R Cygni	7.5	July 4
U Cygni	7.2	July 31
RT Cygni	7.3	Feb. 17, Aug. 26
Chi Cygni	5.2	Nov. 8
R Geminorum	7.1	Nov. 6
U Herculis	7.5	Oct. 12
R Horologii	6.0	Dec. 26
U Horologii	7.0	Jan. 24
R Hydræ	4.5	May 14
R Leonis	5.8	Aug. 20
R Leonis Minoris	7.1	Sep. 12
RS Libræ	7.5	Feb. 23, Sep. 29

Star	*Mean magnitude at maximum*	*Dates of maxima*
V Monocerotis	7.0	Jan. 31
T Normæ	7.4	Apr. 3, Nov. 30
V Ophiuchi	7.5	July 22
X Ophiuchi	6.8	Sep. 13
U Orionis	6.3	Nov. 8
R Sagittarii	7.3	Feb. 25, Nov. 22
RR Sagittarii	6.8	Nov. 20
RT Sagittarii	7.0	June 22
RU Sagittarii	7.2	Aug. 4
RR Scorpii	5.9	July 2
RS Scorpii	7.0	Sep. 9
S Sculptoris	6.7	Dec. 7
R Serpentis	6.9	Feb. 6
R Trianguli	6.2	Jan. 13, Oct. 7
R Ursæ Majoris	7.5	July 3
R Virginis	6.9	Jan. 3, May 29, Oct. 21
S Virginis	7.0	Oct. 24

Some Interesting Double Stars

R. W. ARGYLE

The positions given below correspond to epoch 1998.0.

Star	RA h	RA m	Declination °	Declination ′	Magnitudes	Separation arc seconds	PA °	Catalogue	Comments
β Tuc	00	31.5	−62	58	4.4, 4.8	27.1	170	L119	Both again difficult doubles.
η Cas	00	49.1	+57	49	3.4, 7.5	12.8	316	Σ60	Easy. Creamy, bluish.
β Phe	01	06.1	−46	43	4.0, 4.2	1.5	324	Slr1	Binary. Yellow. Slowly closing.
ζ Psc	01	13.7	+07	35	5.6, 6.5	22.8	63	Σ100	Yellow, reddish-white.
p Eri	01	39.8	−56	12	5.8, 5.8	11.5	191	Δ5	Period 483 years.
γ Ari	01	53.5	+19	18	4.8, 4.8	7.6	0	Σ180	Very easy. Both white.
α Psc	02	02.0	+02	46	4.2, 5.1	1.8	273	Σ202	Period 933 years.
γ And	02	03.9	+42	20	2.3, 5.0	9.4	64	Σ205	Yellow, blue. B is double. Needs 30 cm.
ι Cas AB	02	29.1	+67	24	4.9, 6.9	2.5	231	Σ262	AB is long-period binary. Period 840 years.
ι Cas AC					4.9, 8.4	7.2	118		
ω For	02	33.8	−28	14	5.0, 7.7	24.5	109	h3506	Common proper motion.
γ Cet	02	43.3	+03	14	3.5, 7.3	2.9	294	Σ299	Not too easy.
θ Eri	02	58.3	−40	18	3.4, 4.5	8.3	90	Pz 2	Both white.
ϵ Ari	02	59.2	+21	20	5.2, 5.5	1.5	208	Σ333	Binary. Both white.
Σ331 Per	03	00.9	+47	41	5.3, 6.7	8.5	123	–	Fixed.
α For	03	12.1	−28	59	4.0, 7.0	5.0	299	h3555	Period 314 years. B variable?
f Eri	03	48.6	−37	37	4.8, 5.3	8.2	215	Δ16	Pale yellow. Fixed.
32 Eri	03	54.3	−02	57	4.8, 6.1	6.8	347	Σ470	Fixed.
1 Cam	04	32.0	+53	55	5.7, 6.8	10.3	308	Σ550	Fixed.
ι Pic	04	50.9	−53	28	5.6, 6.4	12.5	58	Δ18	Good object for small apertures. Fixed.
ϰ Lep	05	13.2	−12	56	4.5, 7.4	2.2	356	Σ661	Visible in 7.5 cm.
β Ori	05	14.5	−08	12	0.1, 6.8	9.5	202	Σ668	Companion once thought to be close double.
41 Lep	05	21.8	−24	46	5.4, 6.6	3.5	92	h3752	Deep yellow pair in a rich field.
η Ori	05	24.5	−02	24	3.8, 4.8	1.9	77	Da5	Slow-moving binary.
λ Ori	05	35.1	+09	56	3.6, 5.5	4.4	43	Σ738	Fixed.

Star	*RA* h m	*Declination* ° ′	*Magnitudes*	*Separation arc seconds*	*PA* °	*Catalogue*	*Comments*
θ Ori AB	05 35.3	−05 23	6.7, 7.9	8.7	32	Σ748	Trapezium in M42.
θ Ori CD			5.1, 6.7	13.4	61		
σ Ori AC	05 38.7	−02 36	4.0, 10.3	11.4	238	Σ762	Quintuple. A is a close double.
σ Ori ED			6.5, 7.5	30.1	231		
ζ Ori	05 40.8	−01 37	1.9, 4.0	2.4	162	Σ774	Can be split in 7.5 cm.
η Gem	06 14.9	+22 30	var., 6.5	1.6	257	β1008	Well seen with 20 cm. Primary orange.
12 Lyn AB	06 46.2	+59 27	5.4, 6.0	1.7	72	Σ948	AB is binary. Period 706 years.
12 Lyn AC			5.4, 7.3	8.6	309		
γ Vol	07 08.8	−70 30	3.9, 5.8	13.8	299	Δ42	Very slow binary.
h3945 CMa	07 16.6	−23 19	4.8, 6.8	26.7	52	–	Contrasting colours.
δ Gem	07 20.1	+21 59	3.5, 8.2	5.8	225	Σ1066	Not too easy. Yellow, pale blue.
α Gem	07 34.6	+31 53	1.9, 2.9	3.7	67	Σ1110	Widening. Easy with 7.5 cm.
ϰ Pup	07 38.8	−26 48	4.5, 4.7	9.8	318	H III 27	Both white.
ζ Cnc AB	08 12.2	+17 39	5.6, 6.0	0.8	99	Σ1196	A is a close double.
ζ Cnc AB-C			5.0, 6.2	5.9	74		
δ Vel	08 44.7	−54 43	2.1, 5.1	2.0	140	I10	Slowly closing. Visible in 10 cm.
ϵ Hyd	08 46.8	+06 25	3.3, 6.8	2.7	300	Σ1273	PA slowly increasing.
38 Lyn	09 18.8	+36 48	3.9, 6.6	2.6	226	Σ1338	Almost fixed.
υ Car	09 47.1	−65 04	3.1, 6.1	5.0	127	Rmk11	Fixed. Fine in small telescopes.
γ Leo	10 20.0	+19 51	2.2, 3.5	4.4	125	Σ1424	Binary. Period 619 years. Both orange.
s Vel	10 31.9	−45 04	6.2, 6.5	13.5	218	Pz	Fixed.
μ Vel	10 39.3	−55 36	2.7, 6.4	2.6	57	R155	Period 116 years. Widest in 1996.
54 Leo	10 55.6	+24 45	4.5, 6.3	6.5	112	Σ1487	Slowly widening. Pale yellow and white.
ξ UMa	11 18.2	+31 32	4.3, 4.8	1.6	287	Σ1523	Binary, 60 years. Opening. Needs 10 cm.
π Cen	11 21.0	−54 29	4.3, 5.0	0.3	140	I879	Binary, 39.2 years. Very close. Needs 35 cm.
ι Leo	11 23.9	+10 32	4.0, 6.7	1.7	119	Σ1536	Period 192 years.
N Hya	11 32.3	−29 16	5.8, 5.9	9.3	210	H III 96	Fixed.
D Cen	12 14.0	−45 43	5.6, 6.8	2.9	244	Rmk14	Orange and white. Closing.
α Cru	12 26.6	−63 06	1.4, 1.9	4.2	114	–	Third star in a low-power field.
γ Cen	12 41.5	−48 58	2.9, 2.9	1.1	349	h4539	Period 84 years. Closing. Both yellow.
γ Vir	12 41.7	−01 27	3.5, 3.5	1.9	272	Σ1670	Period 168 years.
β Mus	12 46.7	−68 06	3.7, 4.0	1.3	41	R207	Both white. Closing.
μ Cru	12 54.6	−57 11	4.3, 5.3	34.9	17	Δ126	Fixed. Both white.
α CVn	12 56.0	+38 19	2.9, 5.5	19.6	228	Σ1692	Easy. Yellow, bluish.

Star	*RA* h m	*Declination* ° ′	*Magnitudes*	*Separation arc seconds*	*PA* °	*Catalogue*	*Comments*
J Cen	13 22.6	−60 59	4.6, 6.5	6.0	343	Δ133	Fixed. A is a close pair.
ζ UMa	13 23.9	+54 56	2.3, 4.0	14.4	151	Σ1744	Very easy. Naked-eye pair with Alcor.
3 Cen	13 51.8	−33 00	4.5, 6.0	7.8	105	H III 101	Both white. Closing slowly.
α Cen	14 39.6	−60 50	0.0, 1.2	15.5	220	Rich	Finest pair in the sky. Period 80 years. Closing.
ζ Boo	14 41.1	+13 44	4.5, 4.6	0.8	300	Σ1865	Both white. Closing, highly inclined orbit.
ϵ Boo	14 45.0	+27 04	2.5, 4.9	2.8	342	Σ1877	Yellow, blue. Fine pair.
54 Hya	14 46.0	−25 27	5.1, 7.1	8.2	123	H III 97	Closing slowly.
μ Lib	14 49.3	−14 09	5.8, 6.7	2.0	5	β106	Becoming wider. Fine in 7.5 cm.
ξ Boo	14 51.4	+19 06	4.7, 7.0	6.7	319	Σ1888	Fine contrast. Easy.
44 Boo	15 03.8	+47 39	5.3, 6.2	2.1	52	Σ1909	Period 246 years.
π Lup	15 05.1	−47 03	4.6, 4.7	1.7	65	h4728	Widening.
μ Lup	15 18.5	−47 53	5.1, 5.2	24.0	129	Δ180	Almost fixed.
γ Cir	15 23.4	−59 19	5.1, 5.5	0.7	10	h4787	Closing. Needs 20 cm. Long-period binary.
η CrB	15 32.0	+32 17	5.6, 5.9	0.9	53	Σ1937	Both yellow. Period 41 years, near widest separation.
δ Ser	15 34.8	+10 32	4.2, 5.2	4.3	176	Σ1954	Long-period binary.
γ Lup	15 35.1	−41 10	3.5, 3.6	0.7	274	h4786	Period 147 years. Needs 20 cm.
ξ Lup	15 56.9	−33 58	5.3, 5.8	13.4	49	Pz	Fixed.
σ CrB	16 14.7	+33 52	5.6, 6.6	7.0	237	Σ2032	Long-period binary. Both white.
α Sco	16 29.4	−26 26	1.2, 5.4	2.7	274	–	Red, green. Difficult from mid-northern latitudes.
λ Oph	16 30.9	+01 59	4.2, 5.2	1.4	27	Σ2055	Period 129 years. Fairly difficult in small apertures.
ζ Her	16 41.3	+31 36	2.9, 5.5	1.1	39	Σ2084	Fine, rapid binary. Period 34 years.
μ Dra	17 05.3	+54 28	5.7, 5.7	2.2	20	Σ2130	Long-period binary. Slowly closing.
α Her	17 14.6	+14 23	var., 5.4	4.6	106	Σ2140	Red, green. Binary.
36 Oph	17 15.3	−26 36	5.1, 5.1	4.9	148	SH243	Period 549 years.
ρ Her	17 23.7	+37 09	4.6, 5.6	4.2	318	Σ2161	Slowly widening.
95 Her	18 01.5	+21 36	5.0, 5.1	6.3	258	Σ2264	Colours thought variable in C19.
70 Oph	18 05.5	+02 30	4.2, 6.0	3.2	155	Σ2272	Opening. Easy in 7.5 cm.
h5014 CrA	18 06.8	−43 25	5.7, 5.7	1.4	5	–	Closing slowly. Orbit poorly known. Needs 10 cm.
OΣ358 Her	18 35.9	+16 59	6.8, 7.0	1.3	149	–	Period 292 years.

Star	*RA* h m	*Declination* ° ′	*Magnitudes*	*Separation arc seconds*	*PA* °	*Catalogue*	*Comments*
ε¹ Lyr	18 44.3	+39 40	5.0, 6.1	2.6	351	Σ2382	Quadruple system with ε². Both pairs visible in 7.5 cm.
ε² Lyr	18 44.3	+39 40	5.2, 5.5	2.3	83	Σ2383	
θ Ser	18 56.2	+04 12	4.5, 5.4	22.3	103	Σ2417	Fixed. Very easy.
γ CrA	19 06.4	−37 04	4.8, 5.1	1.3	66	h5084	Beautiful pair. Period 122 years.
β Cyg AB	19 30.7	+27 57	3.1, 5.1	34.1	54	ΣI 43	Glorious. Yellow, blue-greenish. Aa closing.
β Cyg Aa			3.1, 4.0	0.4	134		
δ Cyg	19 45.0	+45 08	2.9, 6.3	2.5	223	Σ2579	Slowly widening.
ε Dra	19 48.2	+70 16	3.8, 7.4	3.2	19	Σ2603	Slow binary.
γ Del	20 46.7	+16 07	4.5, 5.5	9.3	267	Σ2727	Easy. Yellowish.
λ Cyg	20 47.4	+36 29	4.8, 6.1	0.8	7	OΣ413	Difficult binary in small apertures.
ε Equ AB	20 59.1	+04 18	6.0, 6.3	0.9	285	Σ2737	Fine triple. AB is closing.
ε Equ AC			6.0, 7.1	10.0	66		
61 Cyg	21 06.9	+38 45	5.2, 6.0	30.5	149	Σ2758	Nearby binary. Both orange. Period 722 years.
θ Ind	21 19.9	−53 27	4.5, 7.0	6.7	266	–	Pale yellow and reddish. Long-period binary.
μ Cyg	21 44.1	+28 45	4.8, 6.1	1.9	306	Σ2822	Period 713 years.
ξ Cep	22 03.8	+64 38	4.4, 6.5	8.1	276	Σ2863	White and blue.
53 Aqr	22 26.6	−16 45	6.4, 6.6	1.7	359	Sh345	Long-period binary. Closing.
ζ Aqr	22 28.8	−00 01	4.3, 4.5	2.1	192	Σ2909	Slowly widening.
Σ3050 And	23 59.5	+33 43	6.6, 6.6	1.8	329	–	Period 355 years.

Some Interesting Nebulae, Clusters and Galaxies

Object	RA h	m	Declination °	′	Remarks
M.31 Andromedæ	00	40.7	+41	05	Great Galaxy, visible to naked eye.
H.VIII 78 Cassiopeiæ	00	41.3	+61	36	Fine cluster, between Gamma and Kappa Cassiopeiæ.
M.33 Trianguli	01	31.8	+30	28	Spiral. Difficult with small apertures.
H.VI 33–4 Persei	02	18.3	+56	59	Double cluster; Sword-handle.
△142 Doradus	05	39.1	−69	09	Looped nebula round 30 Doradus. Naked-eye. In Large Magellanic Cloud.
M.1 Tauri	05	32.3	+22	00	Crab Nebula, near Zeta Tauri.
M.42 Orionis	05	33.4	−05	24	Great Nebula. Contains the famous Trapezium, Theta Orionis.
M.35 Geminorum	06	06.5	+24	21	Open cluster near Eta Geminorum.
H.VII 2 Monocerotis	06	30.7	+04	53	Open cluster, just visible to naked eye.
M.41 Canis Majoris	06	45.5	−20	42	Open cluster, just visible to naked eye.
M.47 Puppis	07	34.3	−14	22	Mag. 5.2. Loose cluster.
H.IV 64 Puppis	07	39.6	−18	05	Bright planetary in rich neighbourhood.
M.46 Puppis	07	39.5	−14	42	Open cluster.
M.44 Cancri	08	38	+20	07	Præsepe. Open cluster near Delta Cancri. Visible to naked eye.
M.97 Ursæ Majoris	11	12.6	+55	13	Owl Nebula, diameter 3′. Planetary.
Kappa Crucis	12	50.7	−60	05	'Jewel Box'; open cluster, with stars of contrasting colours.
M.3 Can. Ven.	13	40.6	+28	34	Bright globular.
Omega Centauri	13	23.7	−47	03	Finest of all globulars. Easy with naked eye.
M.80 Scorpii	16	14.9	−22	53	Globular, between Antares and Beta Scorpii.
M.4 Scorpii	16	21.5	−26	26	Open cluster close to Antares.
M.13 Herculis	16	40	+36	31	Globular. Just visible to naked eye.
M.92 Herculis	16	16.1	+43	11	Globular. Between Iota and Eta Herculis.
M.6 Scorpii	17	36.8	−32	11	Open cluster; naked eye.
M.7 Scorpii	17	50.6	−34	48	Very bright open cluster; naked eye.
M.23 Sagittarii	17	54.8	−19	01	Open cluster nearly 50′ in diameter.
H.IV 37 Draconis	17	58.6	+66	38	Bright planetary.
M.8 Sagittarii	18	01.4	−24	23	Lagoon Nebula. Gaseous. Just visible with naked eye.
NGC 6572 Ophiuchi	18	10.9	+06	50	Bright planetary, between Beta Ophiuchi and Zeta Aquilæ.
M.17 Sagittarii	18	18.8	−16	12	Omega Nebula. Gaseous. Large and bright.
M.11 Scuti	18	49.0	−06	19	Wild Duck. Bright open cluster.
M.57 Lyræ	18	52.6	+32	59	Ring Nebula. Brightest of planetaries.
M.27 Vulpeculæ	19	58.1	+22	37	Dumbbell Nebula, near Gamma Sagittæ.
H.IV 1 Aquarii	21	02.1	−11	31	Bright planetary near Nu Aquarii.
M.15 Pegasi	21	28.3	+12	01	Bright globular, near Epsilon Pegasi.
M.39 Cygni	21	31.0	+48	17	Open cluster between Deneb and Alpha Lacertæ. Well seen with low powers.

Our Contributors

Richard L. S. Taylor was a lecturer on astronomy, planetary and space sciences for the Extra-Mural Department of London University for more than 25 years. Currently the Scientific Secretary of a specialist research group, he is involved with the study of Mars and the possibility that the planet may ultimately be colonized by humans.

Dr David A. Rothery is a Senior Lecturer in Earth Sciences at the Open University, where he teaches geology and planetary science. His research interests include volcanology on the Earth and other Solar System bodies.

Dr Paul Murdin is the Head of Astronomy at the Particle Physics and Astronomy Research Council and the Director of Science at the British National Space Centre.

Dr Andrew J. Hollis has his observatory at Winsford in Cheshire; he is Director of the Remote Planets Section of the British Astronomical Association.

Dr G. J. H. McCall is a geologist who has specialized in meteoritic research. He was for many years carrying out his researches in Australia, but has now returned to England, and lives in Gloucestershire.

Dr Helen J. Walker, of the Rutherford Appleton Laboratory, is an astrophysicist who specializes in interstellar material.

Professor Chris Kitchin is Director of the University of Hertfordshire Observatory, and is primarily an astrophysicist.

Revd Robert Evans, an Australian clergyman, may well be regarded as the world's foremost amateur astronomer; he makes a habit of discovering supernovæ in external galaxies. He has received many honours, including the Centenary Medal of the Société Astronomique de France and the Medal of the Order of Australia for contributions to science.

Iain Nicolson, formerly of the University of Hertfordshire Observatory, now concentrates upon writing as well as research and has now moved back to his native Scotland. He is one of our most regular and valued contributors.

> The William Herschel Society maintains the museum established at 19 New King Street, Bath – the only surviving Herschel House. It also undertakes activities of various kinds. New members would be welcome; those interested are asked to contact the Secretary at the museum.

Astronomical Societies in the British Isles

British Astronomical Association
Assistant Secretary: Burlington House, Piccadilly, London W1V 9AG.
Meetings: Lecture Hall of Scientific Societies, Civil Service Commission Building, 23 Savile Row, London W1. Last Wednesday each month (Oct.–June). 5 p.m. and some Saturday afternoons.

Association for Astronomy Education
Secretary: Bob Kibble, 34 Ackland Crescent, Denmark Hill, London SE5 8EQ.

Astronomy Ireland
Secretary: Tony Ryan, PO Box 2888, Dublin 1, Ireland.
Meetings: 2nd and 4th Mondays of each month. Telescope meetings, every clear Saturday.

Federation of Astronomical Societies
Secretary: Mrs Christine Sheldon, Whitehaven, Lower Moor, Pershore, Worcs.

Junior Astronomical Society of Ireland
Secretary: K. Nolan, 5 St Patrick's Crescent, Rathcoole, Co. Dublin.
Meetings: The Royal Dublin Society, Ballsbridge, Dublin 4. Monthly.

Aberdeen and District Astronomical Society
Secretary: Ian C. Giddings, 95 Brentfield Circle, Ellon, Aberdeenshire AB41 9DB.
Meetings: Robert Gordon's Institute of Technology, St Andrew's Street, Aberdeen. Friday 7.30 p.m.

Abingdon Astronomical Society (was Fitzharry's Astronomical Society)
Secretary: Chris Holt, 9 Rutherford Close, Abingdon, Oxon OX14 2AT.
Meetings: All Saints' Methodist Church Hall, Dorchester Crescent, Abingdon, Oxon. 2nd Monday each month, 8 p.m.

Altrincham and District Astronomical Society
Secretary: Colin Henshaw, 10 Delamore Road, Gatley, Cheadle, Cheshire.
Meetings: Public Library, Timperley. 1st Friday of each month, 7.30 p.m.

Astra Astronomy Section
Secretary: Ian Downie, 151 Sword Street, Glasgow G31.
Meetings: Public Library, Airdrie. Weekly.

Aylesbury Astronomical Society
Secretary: Alan Smith, 182 Morley Fields, Leighton Buzzard, Bedfordshire LU7 8WN.
Meetings: 1st Monday in month. Details from Secretary.

Bassetlaw Astronomical Society
Secretary: H. Moulson, 5 Magnolia Close, South Anston, South Yorks.
Meetings: Rhodesia Village Hall, Rhodesia, Worksop, Notts. On 2nd and 4th Tuesdays of month at 8 p.m.

Batley & Spenborough Astronomical Society
Secretary: Robert Morton, 22 Links Avenue, Cleckheaton, West Yorks BD19 4EG.
Meetings: Milner K. Ford Observatory, Wilton Park, Batley. Every Thursday, 7.30 p.m.

Bedford Astronomical Society
Secretary: D. Eagle, 24 Copthorne Close, Oakley, Bedford.
Meetings: Bedford School, Burnaby Rd, Bedford. Last Tuesday each month.

Bingham & Brookes Space Organization
Secretary: N. Bingham, 15 Hickmore's Lane, Lindfield, W. Sussex.

Birmingham Astronomical Society
Secretary: J. Spittles, 28 Milverton Road, Knowle, Solihull, West Midlands.
Meetings: Room 146, Aston University, last Tuesday each month, Sept. to June (except Dec., moved to 1st week in Jan.).

Blackpool & District Astronomical Society
Secretary: J. L. Crossley, 24 Fernleigh Close, Bispham, Blackpool, Lancs.

Bolton Astronomical Society
Secretary: Peter Miskiw, 9 Hedley Street, Bolton.

Border Astronomical Society
Secretary: David Pettit, 14 Shap Grove, Carlisle, Cumbria.

Boston Astronomers
Secretary: B. Tongue, South View, Fen Road, Stickford, Boston.
Meetings: Details from the Secretary.

Bradford Astronomical Society
Secretary: Mr Rod Hine, 149 Botton Hall Road, Botton Woods, Bradford BD2 1BQ.
Meetings: Eccleshill Library, Bradford 2. Monday fortnightly (with occasional variations).

Braintree, Halstead & District Astronomical Society
Secretary: Heather Reeder, The Knoll, St Peters in the Field, Braintree, Essex.
Meetings: St Peter's Church Hall, St Peter's Road, Braintree, Essex. 3rd Thursday each month, 8 p.m.

Breckland Astronomical Society (was Great Ellingham and District Astronomy Club)
Secretary: Andrew Briggs, Avondale, Norwich Road, Besthorpe, Norwich NR17 2LB.
Meetings: Great Ellingham Recreation Centre, Watton Road (B1077), Great Ellingham, 2nd Friday each month, 7.15 p.m.

Bridgend Astronomical Society
Secretary: Clive Down, 10 Glan y Llyn, Broadlands, North Cornelly, Bridgend.
Meetings: G.P. Room, Recreation Centre, Bridgend, 1st and 3rd Friday monthly, 7.30 p.m.

Bridgwater Astronomical Society
Secretary: W. L. Buckland, 104 Polden Street, Bridgwater, Somerset.
Meetings: Room D10, Bridgwater College, Bath Road Centre, Bridgwater. 2nd Wednesday each month, Sept.–June.

Brighton Astronomical Society
Secretary: Mrs B. C. Smith, Flat 2, 23 Albany Villas, Hove, Sussex BN3 2RS.
Meetings: Preston Tennis Club, Preston Drive, Brighton. Weekly, Tuesdays.

Bristol Astronomical Society
Secretary: Geoff Cane, 9 Sandringham Road, Stoke Gifford, Bristol.
Meetings: Royal Fort (Rm G44), Bristol University. Every Friday each month, Sept.–May. Fortnightly, June–Aug.

Cambridge Astronomical Association
Secretary: R. J. Greening, 20 Cotts Croft, Great Chishill, Royston, Herts.
Meetings: Venues as published in newsletter. 1st and 3rd Friday each month, 8 p.m.

Cardiff Astronomical Society
Secretary: D. W. S. Powell, 1 Tal-y-Bont Road, Ely, Cardiff.
Meetings: Room 230, Dept. Law, University College, Museum Avenue, Cardiff. Alternate Thursdays, 8 p.m.

Castle Point Astronomy Club
Secretary: Andrew Turner, 3 Canewdon Hall Close, Canewdon, Essex SS4 3PY.
Meetings: St Michael's Church, Thundersley. Wednesdays, 8 p.m.

Chelmsford Astronomers
Secretary: Brendan Clark, 5 Borda Close, Chelmsford, Essex.
Meetings: Once a month.

Chester Astronomical Society
Secretary: Mrs S. Brooks, 39 Halton Road, Great Sutton, South Wirral.
Meetings: Southview Community Centre, Southview Road, Chester. Last Monday each month except Aug. and Dec., 7.30 p.m.

Chester Society of Natural Science Literature and Art
Secretary: Paul Braid, 'White Wing', 38 Bryn Avenue, Old Colwyn, Colwyn Bay, Clwyd.
Meetings: Grosvenor Museum, Chester. Fortnightly.

Chesterfield Astronomical Society
Secretary: P. Lisewski, 148 Old Hall Road, Brampton, Chesterfield.
Meetings: Barnet Observatory, Newbold. Each Friday.

Clacton & District Astronomical Society
Secretary: C. L. Haskell, 105 London Road, Clacton-on-Sea, Essex.

Cleethorpes & District Astronomical Society
Secretary: C. Illingworth, 38 Shaw Drive, Grimsby, S. Humberside.
Meetings: Beacon Hill Observatory, Cleethorpes. 1st Wednesday each month.

Cleveland & Darlington Astronomical Society
Secretary: Neil Haggath, 5 Fountains Crescent, Eston, Middlesbrough, Cleveland.
Meetings: Elmwood Community Centre, Greens Lane, Hartburn, Stockton-on-Tees. Monthly, usually 2nd Friday.

Colchester Amateur Astronomers
Secretary: F. Kelly, 'Middleton', Church Road, Elmstead Market, Colchester, Essex.
Meetings: William Loveless Hall, High Street, Wivenhoe. Friday evenings. Fortnightly.

Cornwall Astronomy Society
Secretary: J. M. Harvey, 2 Helland Gardens, Penryn, Cornwall.
Meetings: Godolphin Club, Wendron Street, Helston, Cornwall. 2nd and 4th Thursday of each month, 7.30 for 8 p.m.

Cotswold Astronomical Society
Secretary: Trevor Talbot, Innisfree, Winchcombe Road, Sedgebarrow, Worcs.
Meetings: Fortnightly in Cheltenham or Gloucester.

Coventry & Warwicks Astronomical Society
Secretary: V. Cooper, 5 Gisburn Close, Woodloes Park, Warwick.
Meetings: Coventry Technical College. 1st Friday each month, Sept.–June.

Crawley Astronomical Society
Secretary: G. Cowley, 67 Climpixy Road, Ifield, Crawley, Sussex.
Meetings: Crawley College of Further Education. Monthly Oct.–June.

Crayford Manor House Astronomical Society
Secretary: R. H. Chambers, Manor House Centre, Crayford, Kent.
Meetings: Manor House Centre, Crayford. Monthly during term-time.

Croydon Astronomical Society
Secretary: John Murrell, 17 Dalmeny Road, Carshalton, Surrey.
Meetings: Lecture Theatre, Royal Russell School, Combe Lane, South Croydon. Alternate Fridays, 7.45 p.m.

Derby & District Astronomical Society
Secretary: Jane D. Kirk, 7 Cromwell Avenue, Findern, Derby.
Meetings: At home of Secretary. 1st and 3rd Friday each month, 7.30 p.m.

Doncaster Astronomical Society
Secretary: J. A. Day, 297 Lonsdale Avenue, Intake, Doncaster.
Meetings: Fridays, weekly.

Dundee Astronomical Society
Secretary: G. Young, 37 Polepark Road, Dundee, Angus.
Meetings: Mills Observatory, Balgay Park, Dundee. 1st Friday each month, 7.30 p.m. Sept.–April.

Easington and District Astronomical Society
Secretary: T. Bradley, 52 Jameson Road, Hartlepool, Co. Durham.
Meetings: Easington Comprehensive School, Easington Colliery. Every 3rd Thursday throughout the year, 7.30 p.m.

Eastbourne Astronomical Society
Secretary: D. C. Gates, Apple Tree Cottage, Stunts Green, Hertsmonceux, East Sussex.
Meetings: St Aiden's Church Hall, 1 Whitley Road, Eastbourne. Monthly (except July and Aug.).

East Lancashire Astronomical Society
Secretary: D. Chadwick, 16 Worston Lane, Great Harwood, Blackburn BB6 7TH.
Meetings: As arranged. Monthly.

Astronomical Society of Edinburgh
Secretary: Graham Rule, 105/19 Causewayside, Edinburgh EH9 1QG.
Meetings: City Observatory, Calton Hill, Edinburgh. Monthly.

Edinburgh University Astronomical Society
Secretary: c/o Dept. of Astronomy, Royal Observatory, Blackford Hill, Edinburgh.

Ewell Astronomical Society
Secretary: Edward Hanna, 91 Tennyson Avenue, Motspur Park, Surrey.
Meetings: 1st Friday of each month.

Exeter Astronomical Society
Secretary: Miss J. Corey, 5 Egham Avenue, Topsham Road, Exeter.
Meetings: The Meeting Room, Wynards, Magdalen Street, Exeter. 1st Thursday of month.

Farnham Astronomical Society
Secretary: Laurence Anslow, 14 Wellington Lane, Farnham, Surrey.
Meetings: Church House, Union Road, Farnham. 2nd Monday each month, 7.45 p.m.

Furness Astronomical Society
Secretary: A. Thompson, 52 Ocean Road, Walney Island, Barrow-in-Furness, Cumbria.
Meetings: St Mary's Church Centre, Dalton-in-Furness. 2nd Saturday in month, 7.30 p.m. Not Aug.

Fylde Astronomical Society
Secretary: 28 Belvedere Road, Thornton, Lancs.
Meetings: Stanley Hall, Rossendale Avenue South. 1st Wednesday each month.

Astronomical Society of Glasgow
Secretary: Malcolm Kennedy, 32 Cedar Road, Cumbernauld, Glasgow.
Meetings: University of Strathclyde, George St., Glasgow. 3rd Thursday each month, Sept.–April.

Greenock Astronomical Society
Secretary: Carl Hempsey, 49 Brisbane Street, Greenock.
Meetings: Greenock Arts Guild, 3 Campbell Street, Greenock.

Grimsby Astronomical Society
Secretary: R. Williams, 14 Richmond Close, Grimsby, South Humberside.
Meetings: Secretary's home. 2nd Thursday each month, 7.30 p.m.

Guernsey: La Société Guernesiaise Astronomy Section
Secretary: G. Falla, Highcliffe, Avenue Beauvais, Ville du Roi, St Peter's Port, Guernsey.
Meetings: The Observatory, St Peter's, Tuesdays, 8 p.m.

Guildford Astronomical Society
Secretary: A. Langmaid, 22 West Mount, Guildford, Surrey.
Meetings: Guildford Institute, Ward Street, Guildford. 1st Thursday each month, except July and Aug., 7.30 p.m.

Gwynedd Astronomical Society
Secretary: P. J. Curtis, Ael-y-bryn, Malltraeth St Newborough, Anglesey, Gwynedd.
Meetings: Physics Lecture Room, Bangor University. 1st Thursday each month, 7.30 p.m.

The Hampshire Astronomical Group
Secretary: R. F. Dodd, 1 Conifer Close, Cowplain, Waterlooville, Hants.
Meetings: Clanfield Observatory. Each Friday, 7.30 p.m.

Astronomical Society of Haringey
Secretary: Jerry Workman, 33 Arthur Road, Holloway, Islington, London N7 6DS.
Meetings: The Hall of the Good Shepherd, Berwick Road, Wood Green. 3rd Wednesday each month, 8 p.m.

Harrogate Astronomical Society
Secretary: P. Barton, 31 Gordon Avenue, Harrogate, North Yorkshire.
Meetings: Harlow Hill Methodist Church Hall, 121 Otley Road, Harrogate. Last Friday each month.

Hastings and Battle Astronomical Society
Secretary: Mrs Karen Pankhurst, 20 High Bank Close, Ore, Hastings, E. Sussex.
Meetings: Details from Secretary.

Havering Astronomical Society
Secretary: Frances Ridgley, 133 Severn Drive, Upminster, Essex RM14 1PP.
Meetings: Cranham Community Centre, Marlborough Gardens, Upminster, Essex. 3rd Monday each month (except July and Aug.).

Heart of England Astronomical Society
Secretary: Jean Poyner, 67 Ellerton Road, Kingstanding, Birmingham B44 0QE.
Meetings: Furnace End Village, every Thursday.

Hebden Bridge Literary & Scientific Society, Astronomical Section
Secretary: F. Parker, 48 Caldene Avenue, Mytholmroyd, Hebden Bridge, West Yorkshire.

Herschel Astronomy Society
Secretary: D. R. Whittaker, 149 Farnham Lane, Slough.
Meetings: Eton College, 2nd Friday each month.

Highlands Astronomical Society
Secretary: Richard Pearce, 1 Forsyth Street, Hopeman, Elgin.
Meetings: The Spectrum Centre, Inverness. 1st Tuesday each month, 7.30 p.m.

Horsham Astronomy Group (was Forest Astronomical Society)
Chairman: Tony Beale, 8 Mill Lane, Lower Beeding, West Sussex.
Meetings: 1st Wednesday each month. For location contact chairman.

Howards Astronomy Club
Secretary: H. Ilett, 22 St Georges Avenue, Warblington, Havant, Hants.
Meetings: To be notified.

Huddersfield Astronomical and Philosophical Society
Secretary: R. A. Williams, 43 Oaklands Drive, Dalton, Huddersfield.
Meetings: 4a Railway Street, Huddersfield. Every Friday, 7.30 p.m.

Hull and East Riding Astronomical Society
Secretary: A. G. Scaife, 19 Beech Road, Elloughton, East Yorks.
Meetings: Wyke 6th Form College, Bricknell Avenue, Hull. 1st and 3rd Wednesday each month, Oct.–Apr., 7.30 p.m.

Ilkeston & District Astronomical Society
Secretary: Trevor Smith, 129 Heanor Road, Smalley, Derbyshire.
Meetings: The Friends Meeting Room, Ilkeston Museum, Ilkeston. 2nd Tuesday monthly, 7.30 p.m.

Ipswich, Orwell Astronomical Society
Secretary: R. Gooding, 168 Ashcroft Road, Ipswich.
Meetings: Orwell Park Observatory, Nacton, Ipswich. Wednesdays 8 p.m.

Irish Astronomical Association
Secretary: Michael Duffy, 26 Ballymurphy Road, Belfast, Northern Ireland.
Meetings: Room 315, Ashby Institute, Stranmills Road, Belfast. Fortnightly. Wednesdays, Sept.–Apr., 7.30 p.m.

Irish Astronomical Society
Secretary: c/o PO Box 2547, Dublin 15, Ireland.

Isle of Man Astronomical Society
Secretary: James Martin, Ballaterson Farm, Peel, Isle of Man IM5 3AB.
Meetings: The Manx Automobile Club, Hill Street, Douglas. 1st Thursday of each month, 8.00 p.m.

Isle of Wight Astronomical Society
Secretary: J. W. Feakins, 1 Hilltop Cottages, High Street, Freshwater, Isle of Wight.
Meetings: Unitarian Church Hall, Newport, Isle of Wight. Monthly.

Keele Astronomical Society
Secretary: Department of Physics, University of Keele, Keele, Staffs.
Meetings: As arranged during term time.

Kettering and District Astronomical Society
Asst. Secretary: Steve Williams, 120 Brickhill Road, Wellingborough, Northants.
Meetings: Quaker Meeting Hall, Northall Street, Kettering, Northants. 1st Tuesday each month. 7.45 p.m.

King's Lynn Amateur Astronomical Association
Secretary: P. Twynman, 17 Poplar Avenue, RAF Marham, King's Lynn.
Meetings: As arranged.

Lancaster and Morecambe Astronomical Society
Secretary: Miss E. Haygarth, 27 Coulston Road, Bowerham, Lancaster.
Meetings: Midland Hotel, Morecambe. 1st Wednesday each month except Jan. 7.30 p.m.

Lancaster University Astronomical Society
Secretary: c/o Students Union, Alexandra Square, University of Lancaster.
Meetings: As arranged.

Laymans Astronomical Society
Secretary: John Evans, 10 Arkwright Walk, The Meadows, Nottingham.
Meetings: The Popular, Bath Street, Ilkeston, Derbyshire. Monthly.

Leeds Astronomical Society
Secretary: A. J. Higgins, 23 Montagu Place, Leeds LS8 2RQ.
Meetings: Lecture Room, City Museum Library, The Headrow, Leeds.

Leicester Astronomical Society
Secretary: Ann Borell, 53 Warden's Walk, Leicester Forest East, Leics.
Meetings: Judgemeadow Community College, Marydene Drive, Evington, Leicester. 2nd and 4th Tuesdays each month, 7.30 p.m.

Letchworth and District Astronomical Society
Secretary: Eric Hutton, 14 Folly Close, Hitchin, Herts.
Meetings: As arranged.

Limerick Astronomy Club
Secretary: Tony O'Hanlon, 26 Ballycannon Heights, Meelick, Co. Clare, Ireland.
Meetings: Limerick Senior College, Limerick, Ireland. Monthly (except June and August), 8 p.m.

Lincoln Astronomical Society
Secretary: G. Winstanley, 36 Cambridge Drive, Washingborough, Lincoln.
Meetings: The Lecture Hall, off Westcliffe Street, Lincoln. 1st Tuesday each month.

Liverpool Astronomical Society
Secretary: David Whittle, 17 Sandy Lane, Tuebrook, Liverpool.
Meetings: City Museum, Liverpool. Wednesdays and Fridays, monthly.

Loughton Astronomical Society
Secretary: Dave Gill, 4 Tower Road, Epping, Essex.
Meetings: Epping Forest College, Borders Lane, Loughton, Essex. Thursdays 8 p.m.

Lowestoft and Great Yarmouth Regional Astronomers (LYRA) Society
Secretary: R. Cheek, 7 The Glades, Lowestoft, Suffolk.
Meetings: Community Wing, Kirkley High School, Kirkley Run, Lowestoft. 3rd Thursday, Sept.–May. Afterwards in School Observatory. 7.15 p.m.

Luton & District Astronomical Society
Secretary: D. Childs, 6 Greenways, Stopsley, Luton.
Meetings: Luton College of Higher Education, Park Square, Luton. Second and last Friday each month, 7.30 p.m.

Lytham St Annes Astronomical Association
Secretary: K. J. Porter, 141 Blackpool Road, Ansdell, Lytham St Annes, Lancs.
Meetings: College of Further Education, Clifton Drive South, Lytham St Annes. 2nd Wednesday monthly Oct.–June.

Macclesfield Astronomical Society
Secretary: Mrs C. Moss, 27 Westminster Road, Macclesfield, Cheshire.
Meetings: The Planetarium, Jodrell Bank, 1st Tuesday each month.

Maidenhead Astronomical Society
Secretary: c/o Chairman, Peter Hunt, Hightrees, Holyport Road, Bray, Berks.
Meetings: Library. Monthly (except July) 1st Friday.

Maidstone Astronomical Society
Secretary: Stephen James, 4 The Cherry Orchard, Haddow, Tonbridge, Kent.
Meetings: Nettlestead Village Hall, 1st Tuesday in month except July and Aug. 7.30 p.m.

Manchester Astronomical Society
Secretary: J. H. Davidson, Godlee Observatory, UMIST, Sackville Street, Manchester 1.
Meetings: At the Observatory, Thursdays, 7.30–9 p.m.

Mansfield and Sutton Astronomical Society
Secretary: G. W. Shepherd, Sherwood Observatory, Coxmoor Road, Sutton-in-Ashfield, Notts.
Meetings: Sherwood Observatory, Coxmoor Road. Last Tuesday each month, 7.45 p.m.

Mexborough and Swinton Astronomical Society
Secretary: Mark R. Benton, 61 The Lea, Swinton, Mexborough, Yorks.
Meetings: Methodist Hall, Piccadilly Road, Swinton, Near Mexborough. Thursdays, 7 p.m.

Mid-Kent Astronomical Society
Secretary: Peter Bassett, 167 Shakespeare Road, Gillingham, Kent.
Meetings: Venue to be arranged. 2nd and last Friday in month.

Milton Keynes Astronomical Society
Secretary: Mike Leggett, 19 Matilda Gardens, Shenley Church End, Milton Keynes MK5 6HT.
Meetings: Rectory Cottage, Bletchley. Alternate Tuesdays.

Moray Astronomical Society
Secretary: Richard Pearce, 1 Forsyth Street, Hopeman, Elgin, Moray, Scotland.
Meetings: Village Hall Close, Co. Elgin.

Newbury Amateur Astronomical Society
Secretary: Mrs A. Davies, 11 Sedgfield Road, Greenham, Newbury, Berks.
Meetings: United Reform Church Hall, Cromwell Road, Newbury. Last Friday of month, Aug.–May.

Newcastle-on-Tyne Astronomical Society
Secretary: C. E. Willits, 24 Acomb Avenue, Seaton Delaval, Tyne and Wear.
Meetings: Zoology Lecture Theatre, Newcastle University. Monthly.

North Aston Space & Astronomical Club
Secretary: W. R. Chadburn, 14 Oakdale Road, North Aston, Sheffield.
Meetings: To be notified.

Northamptonshire Natural History Astronomical Society
Secretary: Dr Nick Hewitt, 4 Daimler Close, Northampton.
Meetings: Humphrey Rooms, Castillian Terrace, Northampton. 2nd and last Monday each month.

North Devon Astronomical Society
Secretary: P. G. Vickery, 12 Broad Park Crescent, Ilfracombe, North Devon.
Meetings: Pilton Community College, Chaddiford Lane, Barnstaple. 1st Wednesday each month, Sept.–May.

North Dorset Astronomical Society
Secretary: J. E. M. Coward, The Pharmacy, Stalbridge, Dorset.
Meetings: Charterhay, Stourton, Caundle, Dorset. 2nd Wednesday each month.

North Gwent Astronomical Society
Secretary: J. Powell, 14 Lancaster Drive, Gilwern, nr Abergavenny, Gwent NP7 0AA
Meetings: Gilwern Community Centre, 15th of each month, 7.30 p.m.

North Staffordshire Astronomical Society
Secretary: N. Oldham, 25 Linley Grove, Alsager, Stoke-on-Trent.
Meetings: 1st Wednesday of each month at Cartwright House, Broad Street, Hanley.

North Western Association of Variable Star Observers
Secretary: Jeremy Bullivant, 2 Beaminster Road, Heaton Mersey, Stockport, Cheshire.
Meetings: Four annually.

Norwich Astronomical Society
Secretary: Malcolm Jones, Tabor House, Norwich Road, Malbarton, Norwich.
Meetings: The Observatory, Colney Lane, Colney, Norwich. Every Friday, 7.30 p.m.

Nottingham Astronomical Society
Secretary: C. Brennan, 40 Swindon Close, Giltbrook, Nottingham.

Oldham Astronomical Society
Secretary: P. J. Collins, 25 Park Crescent, Chadderton, Oldham.
Meetings: Werneth Park Study Centre, Frederick Street, Oldham. Fortnightly, Friday.

Open University Astronomical Society
Secretary: Jim Lee, c/o above, Milton Keynes.
Meetings: Open University, Walton Hall, Milton Keynes. As arranged.

Orpington Astronomical Society
Secretary: Dr Ian Carstairs, 38 Brabourne Rise, Beckenham, Kent BR3 2SG.
Meetings: Orpington Parish Church Hall, Bark Hart Road. Thursdays monthly, 7.30 p.m. Sept.–July.

Peterborough Astronomical Society
Secretary: Sheila Thorpe, 6 Cypress Close, Longthorpe, Peterborough.
Meetings: 1st Thursday every month at 7.30 p.m.

Plymouth Astronomical Society
Secretary: Sheila Evans, 40 Billington Close, Eggbuckland, Plymouth.
Meetings: Glynnis Kingdon Centre. 2nd Friday each month.

Port Talbot Astronomical Society (was Astronomical Society of Wales)
Secretary: J. A. Minopoli, 11 Tan Y Bryn Terrace, Penclowdd, Swansea.
Meetings: Port Talbot Arts Centre, 1st Tuesday each month, 7.15 p.m.

Portsmouth Astronomical Society
Secretary: G. B. Bryant, 81 Ringwood Road, Southsea.
Meetings: Monday. Fortnightly.

Preston & District Astronomical Society
Secretary: P. Sloane, 77 Ribby Road, Wrea Green, Kirkham, Preston, Lancs.
Meetings: Moor Park (Jeremiah Horrocks) Observatory, Preston. 2nd Wednesday, last Friday each month. 7.30 p.m.

The Pulsar Group
Secretary: Barry Smith, 157 Reridge Road, Blackburn, Lancs.
Meetings: Amateur Astronomy Centre, Clough Bank, Bacup Road, Todmorden, Lancs. 1st Thursday each month.

Reading Astronomical Society
Secretary: Mrs Muriel Wrigley, 516 Wokingham Road, Earley, Reading.
Meetings: St Peter's Church Hall, Church Road, Earley. Monthly (3rd Saturday), 7 p.m.

Renfrew District Astronomical Society (formerly Paisley A.S.)
Secretary: D. Bankhead, 3c School Wynd, Paisley.
Meetings: Coats Observatory, Oakshaw Street, Paisley. Fridays, 7.30 p.m.

Richmond & Kew Astronomical Society
Secretary: Stewart McLaughlin, 41a Bruce Road, Mitcham, Surrey CR4 2BJ.
Meetings: Richmond Adult College, Parkshot, Richmond, Surrey, and the King's Observatory, Old Deer Park, Richmond, Surrey. Bimonthly.

Rower Astronomical Club
Secretary: Mary Kelly, Knockatore, The Rower, Thomastown, Co. Kilkenny, Ireland.
Salford Astronomical Society
Secretary: J. A. Handford, 45 Burnside Avenue, Salford 6, Lancs.
Meetings: The Observatory, Chaseley Road, Salford.
Salisbury Astronomical Society
Secretary: Mrs R. Collins, Mountains, 3 Fairview Road, Salisbury, Wilts.
Meetings: Salisbury City Library, Market Place, Salisbury.
Sandbach Astronomical Society
Secretary: Phil Benson, 8 Gawsworth Drive, Sandbach, Cheshire.
Meetings: Sandbach School, as arranged.
Scarborough & District Astronomical Society
Secretary: Mrs S. Anderson, Basin House Farm, Sawdon, Scarborough, N. Yorks.
Meetings: Scarborough Public Library. Last Saturday each month, 7–9 p.m.
Scottish Astronomers Group
Secretary: G. Young c/o Mills Observatory, Balgay Park, Ancrum, Dundee.
Meetings: Bimonthly, around the country. Syllabus given on request.
Sheffield Astronomical Society
Secretary: Mrs Lilian M. Keen, 21 Seagrave Drive, Gleadless, Sheffield.
Meetings: City Museum, Weston Park, 3rd Friday each month. 7.30 p.m.
Sidmouth and District Astronomical Society
Secretary: M. Grant, Salters Meadow, Sidmouth, Devon.
Meetings: Norman Lockyer Observatory, Salcombe Hill. 1st Monday in each month.
Society for Popular Astronomy (was Junior Astronomical Society)
Secretary: Guy Fennimore, 36 Fairway, Keyworth, Nottingham.
Meetings: Last Saturday in Jan., Apr., July, Oct., 2.30 p.m. in London.
Solent Amateur Astronomers
Secretary: Ken Medway, 443 Burgess Road, Swaythling, Southampton SO16 3BL.
Meetings: Room 2, Oaklands Community Centre, Fairisle Road, Lordshill, Southampton. 3rd Tuesday.
Southampton Astronomical Society
Secretary: M. R. Hobbs, 124 Winchester Road, Southampton.
Meetings: Room 148, Murray Building, Southampton University, 2nd Thursday each month, 7.30 p.m.
South Downs Astronomical Society
Secretary: J. Green, 46 Central Avenue, Bognor Regis, West Sussex.
Meetings: Assembly Rooms, Chichester. 1st Friday in each month.
South-East Essex Astronomical Society
Secretary: C. Jones, 92 Long Riding, Basildon, Essex.
Meetings: Lecture Theatre, Central Library, Victoria Avenue, Southend-on-Sea. Generally 1st Thursday in month, Sept.–May.
South-East Kent Astronomical Society
Secretary: P. Andrew, 7 Farncombe Way, Whitfield, nr. Dover.
Meetings: Monthly.
South Lincolnshire Astronomical & Geophysical Society
Secretary: Ian Farley, 12 West Road, Bourne, Lincs.
Meetings: South Holland Centre, Spalding. 3rd Thursday each month, Sept.–May. 7.30 p.m.
Southport Astronomical Society
Secretary: R. Rawlinson, 188 Haig Avenue, Southport, Merseyside.
Meetings: Monthly Sept.–May, plus observing sessions.
Southport, Ormskirk and District Astronomical Society
Secretary: J. T. Harrison, 92 Cottage Lane, Ormskirk, Lancs L39 3NJ.
Meetings: Saturday evenings, monthly as arranged.
South Shields Astronomical Society
Secretary: c/o South Tyneside College, St George's Avenue, South Shields.
Meetings: Marine and Technical College. Each Thursday, 7.30 p.m.
South Somerset Astronomical Society
Secretary: G. McNelly, 11 Laxton Close, Taunton, Somerset.
Meetings: Victoria Inn, Skittle Alley, East Reach, Taunton. Last Saturday each month, 7.30 p.m.
South-West Cotswolds Astronomical Society
Secretary: C. R. Wiles, Old Castle House, The Triangle, Malmesbury, Wilts.
Meetings: 2nd Friday each month, 8 p.m. (Sept.–June).
South-West Herts Astronomical Society
Secretary: Frank Phillips, 54 Highfield Way, Rickmansworth, Herts.
Meetings: Rickmansworth. Last Friday each month, Sept.–May.
Stafford and District Astronomical Society
Secretary: Mrs L. Hodkinson, Beecholme, Francis Green Lane, Penkridge, Staffs.
Meetings: Riverside Centre, Stafford. Every 3rd Thursday, Sept.–May, 7.30 p.m.

Stirling Astronomical Society
Secretary: Mrs C. Traynor, 5c St Mary's Wynd, Stirling.
Meetings: Smith Museum & Art Gallery, Dumbarton Road, Stirling. 2nd Friday each month, 7.30 p.m.

Stoke-on-Trent Astronomical Society
Secretary: M. Pace, Sundale, Dunnocksfold Road, Alsager, Stoke-on-Trent.
Meetings: Cartwright House, Broad Street, Hanley. Monthly.

Sussex Astronomical Society
Secretary: Mrs C. G. Sutton, 75 Vale Road, Portslade, Sussex.
Meetings: English Language Centre, Third Avenue, Hove. Every Wednesday, 7.30–9.30 p.m. Sept.–May.

Swansea Astronomical Society
Secretary: D. F. Tovey, 43 Cecil Road, Gowerton, Swansea.
Meetings: Lecture Room C, Mathematics and Physics Building, University of Wales, Swansea. 2nd and 4th Thursday each month, 7.00 p.m.

Tavistock Astronomical Society
Secretary: Mrs Ellie Coombes, Rosemount, Under Road, Gunnislake, Cornwall PL18 9JL.
Meetings: Science Laboratory, Kelly College, Tavistock. 1st Wednesday in month. 7.30 p.m.

Thames Valley Astronomical Group
Secretary: K. J. Pallet, 82a Tennyson Street, South Lambeth, London SW8 3TH.
Meetings: As arranged.

Thanet Amateur Astronomical Society
Secretary: P. F. Jordan, 85 Crescent Road, Ramsgate.
Meetings: Hilderstone House, Broadstairs, Kent. Monthly.

Torbay Astronomical Society
Secretary: R. Jones, St Helens, Hermose Road, Teignmouth, Devon.
Meetings: Town Hall, Torquay. 3rd Thursday, Oct.–May.

Tullamore Astronomical Society
Secretary: S. McKenna, 145 Arden Vale, Tullamore, Co. Offaly, Ireland.
Meetings: Tullamore Vocational School. Fortnightly, Tuesdays, Oct.–June. 8 p.m.

Tyrone Astronomical Society
Secretary: John Ryan, 105 Coolnafranky Park, Cookstown, Co. Tyrone.
Meetings: Contact Secretary.

Usk Astronomical Society
Secretary: D. J. T. Thomas, 20 Maryport Street, Usk, Gwent.
Meetings: Usk Adult Education Centre, Maryport Street. Weekly, Thursdays (term dates).

Vectis Astronomical Society
Secretary: J. W. Smith, 27 Forest Road, Winford, Sandown, Isle of Wight.
Meetings: 4th Friday each month, except Dec. at Lord Louis Library Meeting Room, Newport, Isle of Wight.

Vigo Astronomical Society
Secretary: Robert Wilson, 43 Admers Wood, Vigo Village, Meopham, Kent DA13 0SP.
Meetings: Vigo Village Hall, as arranged.

Webb Society
Secretary: M. B. Swan, 194 Foundry Lane, Freemantle, Southampton, Hants.
Meetings: As arranged.

Wellingborough District Astronomical Society
Secretary: S. M. Williams, 120 Brickhill Road, Wellingborough, Northants.
Meetings: On 2nd Wednesday. Gloucester Hall, Church Street, Wellingborough, 7.30 p.m.

Wessex Astronomical Society
Secretary: Leslie Fry, 14 Hanhum Road, Corfe Mullen, Dorset.
Meetings: Allendale Centre, Wimborne, Dorset. 1st Tuesday of each month.

West of London Astronomical Society
Secretary: Tom. H. Ella, 25 Boxtree Road, Harrow Weald, Harrow, Middlesex.
Meetings: Monthly, alternately at Hillingdon and North Harrow. 2nd Monday in month, except Aug.

West Midlands Astronomical Association
Secretary: Miss S. Bundy, 93 Greenridge Road, Handsworth Wood, Birmingham.
Meetings: Dr Johnson House, Bull Street, Birmingham. As arranged.

West Yorkshire Astronomical Society
Secretary: K. Willoughby, 11 Hardisty Drive, Pontefract, Yorks.
Meetings: Rosse Observatory, Carleton Community Centre, Carleton Road, Pontefract, each Tuesday, 7.15 to 9 p.m.

Whitby Astronomical Group
Secretary: Mark Dawson, 33 Laburnum Grove, Whitby, North Yorkshire YO21 1HZ.
Meetings: Mission to Seamen, Haggersgate, Whitby. 2nd Tuesday of the month, 7.30 p.m.

Whittington Astronomical Society
Secretary: Peter Williamson, The Observatory, Top Street, Whittington, Shropshire.
Meetings: The Observatory every month.

Wiltshire Astronomical Society
Secretary: Simon Barnes, 25 Woodcombe, Melksham, Wilts SN12 7SD.
Meetings: St Andrews Church Hall, Church Lane, off Forest Road, Melksham, Wilts.

Wolverhampton Astronomical Society
Secretary: M. Astley, Garwick, 8 Holme Mill, Fordhouses, Wolverhampton.
Meetings: Beckminster Methodist Church Hall, Birches Road, Wolverhampton. Alternate Mondays, Sept.–Apr.

Worcester Astronomical Society
Secretary: Arthur Wilkinson, 179 Henwick Road, St Johns, Worcester.
Meetings: Room 117, Worcester College of Higher Education, Henwick Grove, Worcester. 2nd Thursday each month.

Worthing Astronomical Society
Contact: G. Boots, 101 Ardingly Drive, Worthing, Sussex.
Meetings: Adult Education Centre, Union Place, Worthing, Sussex. 1st Wednesday each month (except Aug.). 7.30 p.m.

Wycombe Astronomical Society
Secretary: P. A. Hodgins, 50 Copners Drive, Holmer Green, High Wycombe, Bucks.
Meetings: 3rd Wednesday each month, 7.45 p.m.

York Astronomical Society
Secretary: Simon Howard, 20 Manor Drive South, Acomb, York.
Meetings: Goddricke College, York University. 1st and 3rd Fridays.

Any society wishing to be included in this list of local societies or to update details is invited to write to the Editor (c/o Macmillan, 25 Eccleston Place, London SW1W 9NF), so that the relevant information may be included in the next edition of the *Yearbook*.